Acclaim for Barbara Strauch's

The Primal Teen

"Upends the longstanding belief that the teenage brain is largely complete, concluding instead that it is undergoing dramatic changes that can help explain what appears to be a gap between intelligence and judgment."
—*The Hartford Courant*

"This is such a smart book. . . . Barbara Strauch acts as a world-class guide to a mysterious place, taking us on a journey through the teenage brain and making sense of the scenery. In turns funny, curious, explanatory, vivid, she does an absolutely compelling job of helping us to understand our children—and ourselves."
—Deborah Blum,
author of *Love at Goon Park: Harry Harlow and the Science of Affection*

"Through interviews with parents, physicians, neuroscientists, and teens, Strauch has compiled impressive insights about the nature of being a teen or the parent of one."
—*Science News*

"Entertaining as well as informative." —*Teacher* magazine

"An intriguing look at cutting-edge studies that now tell us the brain is not finished growing in a child's early years but continues into the teens." —*The Plain Dealer*

"Can knowing more about the teenager's brain help us to understand the teenager's behavior? Can an account of the neuroscience of adolescence be lively and readable? Barbara Strauch provides convincing evidence that the answer to both questions is yes." —Judith Rich Harris, author of *The Nurture Assumption: Why Children Turn Out the Way They Do*

"Readers will be struck by the wonderfully candid comments by those interviewed as well as Strauch's insightful narrative." —*Publishers Weekly*

"Strauch's well-researched book explains studies that were impossible without such advanced technology as the MRI in clear, compassionate, layperson's language. . . . A parents' must-read." —*Booklist*

Barbara Strauch

The Primal Teen

Barbara Strauch is the medical science and health editor of *The New York Times*. She previously covered science and medical issues in Boston and Houston and directed Pulitzer Prize–winning news at *Newsday*. She is the mother of two teenagers and lives in Westchester County, New York.

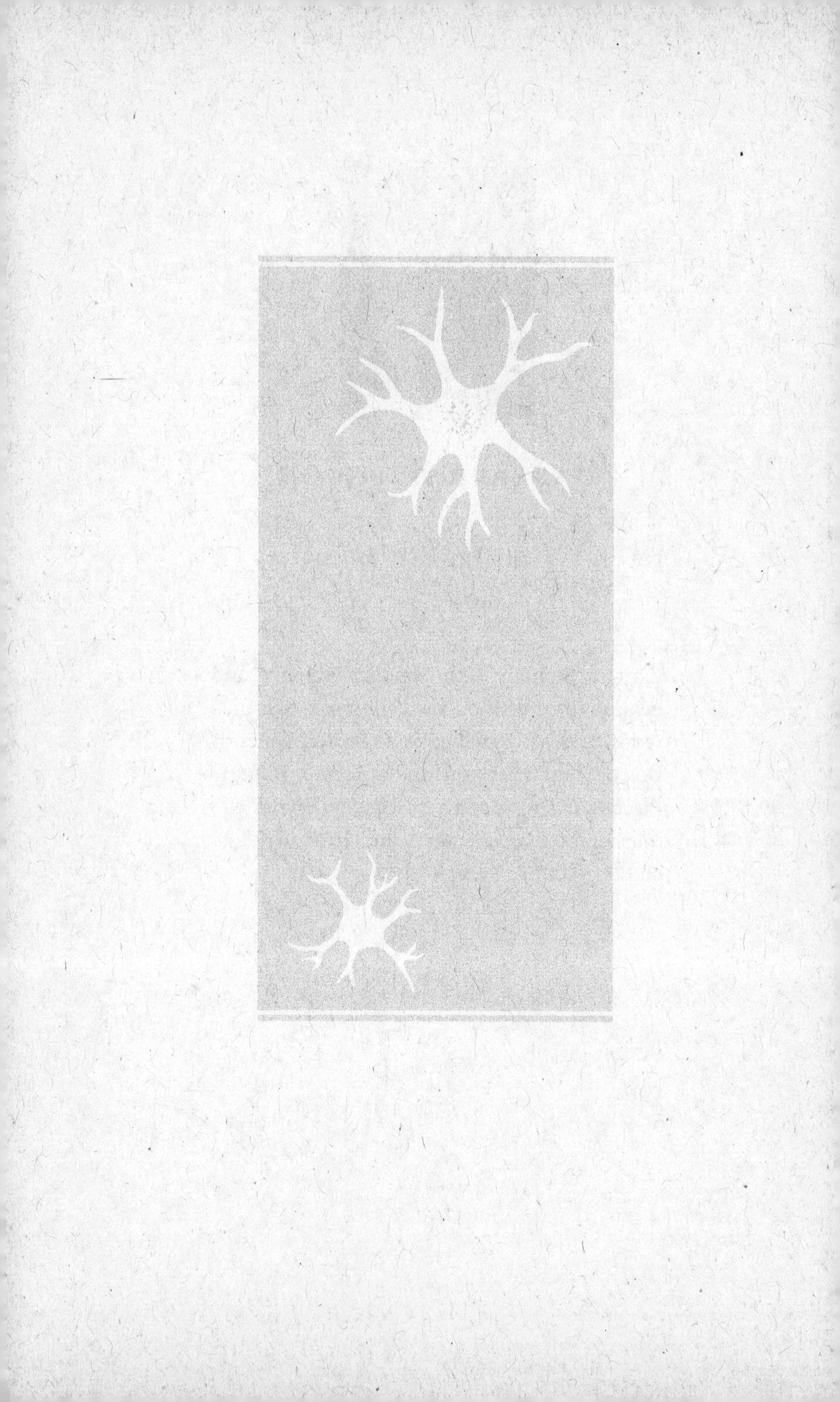

The Primal Teen

What the New Discoveries about the Teenage Brain Tell Us about Our Kids

Barbara Strauch

ANCHOR BOOKS

A Division of Random House, Inc. • New York

To my family—especially the teenagers

FIRST ANCHOR BOOKS EDITION, SEPTEMBER 2004

 Published in the United States by Anchor Books, a division of Random House, Inc., New York, and simultaneously in Canada by Random House of Canada Limited, Toronto. Originally published in hardcover in the United States by Doubleday, a division of Random House, Inc., New York, in 2003.

Ornament by Margaret M. Wagner

The Library of Congress has cataloged the Doubleday edition as follows:
Strauch, Barbara.
The primal teen: what the new discoveries about the teenage brain tell us about our kids / Barbara Strauch.—1st ed.
p. cm.
Includes bibliographical references.
1. Behavioral assessment of teenagers. 2. Teenagers—Physiology. 3. Teenagers—Mental health. 4. Developmental neurobiology. I. Title.
RJ503.S77 2003
616.89'00835—dc21
2002041008

Anchor ISBN: 0-385-72160-9

www.anchorbooks.com

Printed in the United States of America
10 9

Contents

Acknowledgments

THIS BOOK would not have happened without a great deal of help and support and I'm happy to have the chance to say thank-you.

I'm indebted to the many scientists who shared their work and ideas, in particular Jay Giedd of the National Institutes of Health, Charles Nelson at the University of Minnesota, and Paul Thompson and Elizabeth Sowell at UCLA. Others took extra time to provide information and answer my continuing stream of questions, including John Mazziotta at UCLA, Steve Suomi at the National Institutes of Health, Linda Spear at the State University of New York in Binghamton, Francine Benes at McLean Hospital, Marian Diamond at UC Berkeley, David Lewis at the University of Pittsburgh, Patricia Goldman-Rakic at Yale, Marc Breedlove at Michigan State University, William Greenough at the University of Illinois, Elizabeth Cauffman at the University of Pittsburgh, Ted Slotkin at Duke University, and Jill Becker at the University of Michigan.

I am especially grateful to the dozens of teenagers, friends, and parents who generously shared their most intimate stories—both good and bad—with me. Jessica Kovler, Jan Weiss, and Elizabeth Molloy made sure I spoke to as many teenagers as I could. The stories are all true, but in some cases, names have been changed to protect identities.

A number of people graciously read the manuscript—and improved it, including Robin Marx, Karen Pennar, and Barbara Pedley. My friends were kind enough to put up with me while I obsessed on brains, and I got invaluable support and guidance, in particular, from Connie Rosenblum and Jack Schwartz.

I would never have started the book without the early encouragement of Doreen Carvajal and Sandy Blakeslee. It would never have happened without the talents of my agent, Katinka Matson.

I could never have finished the project without the research help provided by Valerie Chris Goelitz or the focused editing of my editor at Doubleday, Roger Scholl, as well as the help from his assistant Sarah Rainone.

I'd also like to thank the entire science department of the *New York Times*, in particular Denise Grady, Gina Kolata, and Erica Goode, as well as Cory Dean, the science editor who graciously agreed to give me time off so I could do research.

Last, I'd like to thank my family for their unwavering patience and understanding. My husband Richard, an editor himself, read and edited the manuscript before I would let anyone else see it—and made me keep working when I didn't want to. My parents, Fred and Claire Strauch, were with me, one way or another, the whole way, as were Ron, Faye, Rowenna, and Nellie.

And most important, I'd like to thank my own two teenagers, Hayley and Meryl, who could not have been more wonderful. They not only shared their thoughts and feelings with me, but they put up with the many days and nights when their mother was behind a closed door—writing about their developing adolescent brains.

Introduction

I MUST SAY, I'd never given much thought to the brains of teenagers. Even as the mother of two teenagers, brains were not the first thing I thought of when trying to figure out why they did what they did. It was hormones; it was friends; it was my fault; it was the summer wind. Brains?

As a science editor, I also knew the official party line about adolescent brains. As far as critical development went, teenage brains were pretty much done, wired up, three pounds of sparkling, charged-up brain cells, just hanging there, ready and waiting for all that Chaucer, all that calculus, all that wise parental caution to pour in.

But not long ago, I began to hear reports from a small group of neuroscientists who had started to look inside the living, working teenage brain, trying to see if they could, in fact, figure out what makes teenagers do what they do.

Their studies had arcane titles: "In Vivo Evidence for Post-Adolescent Brain Maturation in Frontal and Striatal Regions.";

"Structural Maturation of Neural Pathways in Adolescents."

But behind those dry titles was an intriguing story. What are these neuroscientists finding in those teenage brains after all? Could they take us somewhere we've never been? Could they help us figure out what makes a teenager such a delightful and perplexing pain in the neck?

With adolescence blooming all around me at home, I decided to go find out. I crossed the country—at times with my own two teenagers in tow—talking to dozens of neuroscientists and scores of adolescents, trying to figure out what one had to do with the other, if anything. I talked with teenagers on their way to Harvard and teenagers who shot heroin. I talked with parents who wondered what all the fuss was over adolescence and parents who'd been tempted, with a degree of justification, to throw their teenager off a small building. I talked with neuroscientists who peer inside human teenage brains, scientists who spend their days with teenage monkeys, and scientists who slice up the brains of adolescent lab rats.

And, in the end, I had to say, they've got something. The teenage brain—monkey, rat, and human—is different, certainly far different from what I ever thought, far different from what the scientists themselves had thought. In fact, it may be as odd and wacky and weird as teenagers themselves.

This book is the story of those wacky and weird teenage brains. It's also a story of the teenagers who have those brains and the scientists who are, at long last, poking around in them and asking questions: Why do teenagers sleep until noon? Why do they slam those doors, forget to call home, drink themselves silly? Why is it some, quite suddenly it seems, slide into the deep reaches of despair, even the ravages of psychosis, while others, quite suddenly it seems, appreciate the beauty and contours of algebra, the nuance of a subtle joke?

For the first time, neuroscience is wrestling with all that. Using its latest tools, its biggest brains, neuroscience is trying to figure out the answer to one of the oldest questions we have about adolescents: Why are they acting that way? The scientists themselves, excited as they are with what they are finding, acknowledge that they are tiptoeing out on a ledge. It's a science in the making and, like the teenagers themselves, a bit wild. Indeed, the field is changing so fast, I found myself checking dates on any scientific study I looked at. "Oh, 1996, too old."

By and large, scientists—and this book, too—refer to adolescence in its broadest definition. While adolescence certainly includes the more precise biological moment of puberty we most often associate with teenagers, it is not just a single moment but actually a series of stages, many of which are unseen, that begin long before the first breasts sprout, and last long after kids have left for college.

Most periods of brain development have some crucial things going on. The brain, as the scientists will tell you over and over, is the most interactive thing on Earth. Just looking at it changes it; just asking it questions changes it. Your brain is changing now, as you read this.

But most scientists working in this area today think that the changes taking place in the brain during adolescence are so profound, they may rival early childhood as a critical period of development. The teenage brain, far from being ready-made, undergoes a period of surprisingly complex and crucial development.

THIS book is not, for the most part, about troubled teenagers. It does not, ultimately, answer the question of why, particularly in recent years, increasing numbers of young teenagers have decided to pick up semiautomatics and shoot their friends in the high school cafeteria. There are dozens of theories, but, in truth, no one yet knows the answer to that, including me.

The scientists hope one day to learn enough to help teenagers who are troubled and sick. Indeed, many of the scientists probing deep into the teenage brain are driven by the idea that, if their new tools and new science can find out what normal development means, they may figure out what goes wrong in schizophrenia, one of the most devastating illnesses on the planet—and a disease that most often starts in adolescence.

But the first step is to understand how a normal teenage brain grows. What is supposed to happen? What parts are getting bigger? What parts are getting smaller? How does this part connect with that part and when? Can we look in a teenage brain and find, hidden in its folds, some answers to even the most fundamental questions, such as why an otherwise tame and timid teenager goes out one morning and puts a big silver ring in her nose?

As it turns out, yes. As it turns out, teenagers may, indeed, be a bit crazy. But they are crazy according to a primal blueprint; they are crazy by design.

The Primal Teen

Chapter 1

Crazy by Design

The New Science of the Teenage Brain

TEEN. BRAIN. Brain. Teen. The words, I concede, go more tongue in cheek than hand in hand. Utter them out loud—mention even that you're writing a book on the teenage brain—and the jokes come popping out, like pimples before a prom.

"What? They have one?"

"Short book."

Steve, the father of two teenage boys, shook his head and wished me luck. As far as he could tell, the teenage brains he knew had recently and inexplicably gone berserk.

"I don't get it," he said. "All of a sudden strange things are happening with the kids we know. These are good kids, bright kids, but the other day one stole some calculators from the high school and sold them, another can't finish any homework. Getting my own kids out the door in the morning now is this monumental thing. What's going on?"

Denise, a writer and the mother of two lovely teenage boys, was ferocious.

"Teenage brain? I'll tell you teenage brain," she told me one morning. "You know my thirteen-year-old? Well, he went to the school dance, and the rule is you have to stay inside and you can't leave. But he and his friends decided they were imprisoned, and so they went outside and ran in a big circle around the gym. Who knows what they were thinking, but the principal found out and called my husband at home. The next day, we talked and talked; we told him he had to apologize, that he had inconvenienced everyone. But he just couldn't get it; he couldn't get outside himself to understand things from someone else's point of view."

And what was next from this thirteen-year-old, a straight-A kid who never caused a speck of trouble before and who, as his mother says, "used to be so meticulous and quiet and sweet?"

One evening, Denise came home from work and found three letters waiting for her: One said her son had, as expected, been put on the honor roll, the second said he'd made all-county orchestra, and the third said he was being suspended from school after being caught hanging around downtown when he was supposed to be in history class.

"You look at them and sometimes they're so articulate and they seem so grown up and you think they have all the pieces, but sometimes they just don't," said Denise. "I have to say that I believed, smugly, that only irresponsible, self-absorbed jerks had children who got into trouble. But something always happens. It's annoying; it's scary; it can make you crazy."

As the mother of two teenage girls, I don't quibble with this. Annoying stuff happens, scary stuff happens. It can make you crazy.

One morning not long ago my oldest daughter, then fifteen, got up (at noon) and to my delight—without being asked!—cleaned her room and lugged her laundry to the washing machine.

Then, as the rinse cycle whirred and we talked in the living room—I suggested that it wasn't such a great idea to take her new CD player to an amusement park with water rides—she morphed, she mutated. Shooting up off the couch—in full-tilt pubescent display—she stomped upstairs, flinging over her shoulder the battle cry of the teenage warrior goddess, circa 2003: "You suck!"

THE world, of course, has been trying to figure out teenagers for centuries. Their behavior has baffled the best of our thinkers. Aristotle said teenagers appeared "fickle in their desires," which are as "transitory as they are vehement." Shakespeare, Romeo and Juliet aside, described adolescence as largely a time for "getting wenches with child, wronging the ancientry, stealing, fighting." And, if he spoke of their brains at all, which was rare, he dismissed them as "boiled."

The first whiff from those boiled brains can be subdued, even subtle.

Rhona, the mother of two teenagers in New Jersey, knew her daughter Susannah was a teenager simply because she "became embarrassed about the music I had on in the car." Bill, the father of two teenage girls in a small town in New York, says he knew adolescence had arrived when one of his daughters didn't speak to him for a week. Ron, the father of four teenagers in Phoenix, always knew one of his kids had hit adolescence when they started spending a lot of time in the bathroom.

Other times, the whiffs become shock waves.

One father in Minneapolis was aghast when his straitlaced son was caught painting graffiti on the side of a store; a mother in Princeton got a call one spring afternoon from the local police who said her well-behaved fourteen-year-old daughter had stolen a T-shirt from the mall; a mother and father were in tears when their beautiful fifteen-year-old daughter sneaked out her window at midnight to meet a twenty-four-year-old man she'd just met.

For Ellen in New York City, the mother of twin boys, the adolescent wave that swept into her house was abrupt, swift, and "horrible." It arrived about the time her sons were in the seventh grade. "To me it was a return to the terrible twos," she told me. "The tantrums, the stomping of feet, the slamming of doors, the fighting, the name-calling, the animalistic behavior. All of a sudden they got very good at evasive behavior, experts at keeping secrets. Their rooms were a mess, their bodies were clean but they had laundry all over the floor. They became very rude and the self-involvement was breathtaking. It was cosmos-annihilating narcissism, that's what it was."

One son tiptoed out of the family's apartment late at night and somehow got himself to New Jersey with his friends; the other—both boys are quite smart—was caught after he hacked into a university computer system. One—no need to say which—managed to make LSD in the high school science lab.

"Believe me, it was a struggle," said Ellen. "I knew all about how teenagers were supposed to be, but still, it knocked me out."

Teenagers themselves are often surprised at how rapidly their world changes, in all sorts of ways. Lisa, a fourteen-year-old girl from Minneapolis who plays lacrosse and actually likes studying Latin, told me it seemed that "all of a sudden" her moods "were all over the place."

"Sometimes, I just get overwhelmed now," she said. "There's all this friend stuff and school and how I look and my parents. I just go in my room and shut the door. My parents want to talk to me, and I don't mean to be mean, but sometimes I just have to go away and calm down by myself."

This is not to say that all teenagers are alike. Parents often describe at least one child, more if the gods smiled, who just seemed to sail through adolescence. But for most teenagers, blessed with natural curiosity and transversing unknown emotional, physical, and hormonal lands, something usually happens, nothing malicious, nothing illegal, just something.

"I get in trouble a lot more now, but it's for stuff I really didn't mean," said Martin, fifteen. "I forget to call home. I don't know why. I just hang out with friends, and I get involved with that and I forget. Then my parents get really mad, and then I get really mad, and it's a big mess."

WHAT'S GOING ON?

So what's really happening? Why do normal, well-behaved teenagers start pouting in their rooms, sneaking out windows, stomping their feet, or making LSD in the school science lab? For many years, the answer had been simple: hormones, the raging hormones of adolescence.

I remember, not long ago, going to an orientation for my daughter when she was entering sixth grade, and listening to the white-haired principal get a good laugh when he told the crowd of jittery parents sitting in front of him on folding chairs: "Don't worry, we know that all the growth in these middle school years goes on—from the neck down."

He was only partly right. No question, there's unmistakable activity from the neck down in the teenage years. No question, testosterone is there skateboarding down the railings; estrogen is throwing her hips around.

But that's not the whole story.

For the first time now, scientists are starting to look beyond hormones to explain teenage behavior. And they're finding clues in an unexpected place: the teenage brain, the boiled, besmirched—and decidedly above the neck—teenage brain.

For years it was thought that the teenage brain was finished, cooked. Most of the truly important human brain development, scientists believed, was over by the first three years of life. Explaining teenagers has been the job of social scientists, psychologists, psychiatrists, educators, and maybe a priest or two, not neuroscientists. What possible interest could still lurk behind those pierced eyebrows, under that orange-spiked hair?

As it turns out, a whole lot.

In unprecedented work, scientists are discovering exactly how the teenage brain works. Using powerful new brain-scanning machines, peering for the first time into living, working teenage brains, coordinating work across countries and across continents, drawing on pioneering work with adolescent primates and even rats, the neuroscientists are finding that the teenage brain, far from being an innocent bystander to hormonal hijinks, is undergoing a dramatic transformation.

The teenage brain, it's now becoming clear, is still very much a work in progress, a giant construction project. Millions of connections are being hooked up; millions more are swept away. Neurochemicals wash over the teenage brain, giving it a new paint job, a new look, a new chance at life. The teenage brain is raw, vulnerable. It's a brain that's still becoming what it will be.

"We used to think that, if there were brain changes at all in adolescence, they were subtle," Elizabeth Sowell, a neuroscientist at UCLA and one of the country's top researchers of the adolescent brain, told me. "Now we know that those changes are not as subtle as we thought. Every time we look at another set of teenage brains, we find something new."

And this growing group of neuroscientists—in some cases with teenagers of their own they need to figure out—are uncovering clues that can help us all understand why teenagers do what they do, zeroing in on what the normal, average teenage brain might be up to during what one normal, average seventeen-year-old-girl described as her "brief insanity."

The portrait is far from one-dimensional.

The teenage brain may, in fact, be briefly insane. But, scientists say, it is crazy by design. The teenage brain is in flux, maddening and muddled. And that's how it's supposed to be.

And the teenage brain is also wondrous. It's the brains of teenagers, after all, that begin to grapple with our knottiest, most abstract concepts, with honesty and justice. In the neuronal nooks and crannies of their evolving brains, teenagers, for the first time, develop true empathy. They may find themselves, often to their own surprise, happy to stay up until three A.M. to listen to a friend in trouble, worrying about the children in war-torn Afghanistan, or passionately falling in love with the nuances of a poem.

"I love teenagers," said a mother whose two children had successfully navigated their teenage years. "I like their intelligence and their growing ability to think for themselves, to argue intelligently, and get excited by ideas. I like sharing books with them and I like the way they can show me things, like how to use a Palm Pilot. And I like their sense of style."

In their own way, the brain scientists, too, have detected this polysided adolescence, a normal brain evolution that includes moments of mayhem as well as growing precision and passion. Inside the teenagers' brains—smart ones, shy ones, silly ones—they've found, and this word comes from the neuroscientists' own ungainly language, exuberance.

Chapter 2

The Passion Within

Peering into the Living Brain in Search of Normal

Nora loped through the clinic door, her long brown hair with a wide purple streak flowing behind her. In recent years, Nora Berenstain had survived a move from Sacramento to Washington, D.C., her parents' divorce, and a horrific seventh grade when her best friend moved away.

Now sixteen, Nora was confident and instantly likable. She volunteered at NOW headquarters and helped edit the school paper. Wearing faded jeans and a flowered shirt, she flopped her tall, thin body down on a chair and burst out: "How ya doin', Dr. Giedd?"

David and Matthew Goldstein, thirteen-year-old twins, came next, bounding in like matching soccer balls. It was their first time at the clinic and they blurted nervous questions. "This machine, what exactly *is* it going to do, Dr. Giedd?" asked Matthew.

Dr. Jay Giedd, the forty-year-old neuroscientist at the National Institutes of Health (NIH) who they'd come to see, is one of a small group of scientists who have recently discovered that the brains of regular, average teenagers like Nora and David and Matthew are far different from what anyone had imagined.

A child psychiatrist, a neuroscientist, and the father of four young kids himself, Giedd is originally from North Dakota. He has an open, round face, a short red beard, and he grinned broadly as each teenager arrived at his clinic trying, with one of his corny North Dakota jokes, to get them to laugh. "Oh, I love your hair," he told Nora when he saw her purple streak. He patted his own balding head. "I wish I had enough to do that."

Tuesdays are brain-scanning nights at the National Institutes of Health. Giedd, one of the country's top brain scanners, had come to the basement room of Building 10 on the NIH campus, as he had every Tuesday from five to midnight, to engage in his latest passion: Trying to understand the normal teenager.

Teenagers usually come to the big brick buildings of the National Institutes of Health because they're sick. On the third floor of Building 10, there are beds filled with glassy-eyed children diagnosed with schizophrenia. But the teenagers who come down the long, white corridors to the basement room of Building 10 on Tuesday nights are not sick. Purpled-haired and baggy-panted, they're as normal and ordinary and healthy as possible. Volunteered by parents, lured by the $60 per visit they're paid, they arrive and, in the name of science, stick their heads into a big, noisy MRI brain-scanning machine.

Giedd has been putting kids' heads in this machine for ten years, scanning and rescanning the brains of hundreds of children and teenagers. It's the world's first such long-term study of brain development in normal kids and it has produced extraordinary results, the ripples of which extend far beyond any individual teenager and any individual brain and—for parents and neuroscientists alike—are likely to change the way we think about teenagers forever.

One big stumbling block in studying the brains of teenagers has been that relatively few of them die. As far as studying human brains go, it's far easier for scientists to get their hands on the brain of a baby or a grandmother than that of a fourteen-year-old. Studying the development of the adolescent animal brain, too, has not been easy. Unlike in humans, adolescence in animals is often, as one biologist put it, "over in an eye blink," making serious long-term study difficult.

Beyond that, there's been the overreaching issue of boredom. Teenage brains were thought to be largely finished, as far as serious, interesting neural development went. The only big thing happening in adolescence, as far as anyone knew, involved those pesky hormones, hair and pimples, all that.

But new machines that can peer inside living brains—and scientists like Giedd who have decided to study those living teenage brains—have put an end to those thoughts. In the last few years, Giedd and other scientists have found that the adolescent brain undergoes a massive remodeling of its basic structure, in areas that affect everything from logic and language to impulses and intuition.

As neuroscientists, Giedd included, remind us, it's a long way from a brain scan to explaining a teenager who makes LSD in a high school science lab. But what's indisputable is that the teenage brain is not just sitting around, a blob on the bench. And, after years of neglect, the study of the adolescent brain has now become fashionable.

One of the biggest projects aimed at understanding the teenage brain, one launched by various branches of NIH, involves recruiting and scanning five hundred children across the country, carefully matching the racial and socioeconomic mix of America. Another major project, being run in part by the Santa Fe Institute, a research center that wrestles with complex systems from physics to economics, is scanning the brains of babies and teenagers and includes some of the most eminent names in neuroscience. "We wanted to study babies and teenagers because it now seems that those are the times of greatest change in the brain's structure and function," John Mazziotta, a pioneer in brain-scanning at UCLA and one of the heads of the Santa Fe project told me. "And we focus on the times of change because it might be at those times that we can have the most influence."

Research on the teenage brain is just beginning, just peeking its sneakered toes over the skateboard's edge. But already findings are upsetting long-held views not only about the adolescent brain but about brain development in general.

In part, it's a story of technology. Just as the Hubble space telescope is opening a new window on our universe, new machines and even new computer equations have spawned a brand-new discipline of adolescent neuroscience.

And it's a story, too, of the teenagers who are volunteering in this never-before-tried effort to understand them from the inside out.

"We need to study normal teenagers and we need to look at those same normal teenage brains over and over," says Giedd. "How can we ever help kids with problems if we don't know what normal is?"

THE first volunteer on the Tuesday night of my visit was Matthew Goldstein, who walked resolutely toward the giant MRI, or magnetic resonance imaging, machine, a gray plastic cube with a hole in the middle. Once Matthew was inside, the MRI, using radio waves, magnetic fields, and computers, would give Giedd a remarkably clear picture of the inner structure of the thirteen-year-old's brain.

At Giedd's direction, Matthew hopped up onto the machine, his feet dangling over the side. Because it was Matthew's first brain scan, Giedd explained the process step by step.

"Now," said Giedd, speaking in the slow, comforting tones of a child psychiatrist as he spoke to Matthew, "I want you to lie back, and I'm going to strap in your head. It's not going to hurt. You'll be fine. There's a mirror in there where you can see us in the other room. Just think of yourself as an astronaut going out into outer space."

Matthew lay back and slid into the scanner, leaving only his feet sticking out on the slide. His father, Ira Goldstein, still dressed in his gray business suit, was quietly standing by.

"Is it OK if I touch him?" Ira finally asked Giedd tentatively.

Giedd assured him that it was fine, and Ira gently placed his hand on his son's foot. Matthew, teenage brain science volunteer, would be in the scanner for forty-five minutes and, for all that time, his father's hand would not waver.

UNDER CONSTRUCTION

Jay Giedd had first seen hints of the remarkable transformation of the teenage brain five years before. Like a number of neuroscientists who scan the human brain, Giedd often sends brain pictures to the Montreal Neurological Institute, where, with super-fast computerization, scans are turned into rows of numbers corresponding to brain section sizes that are neatly e-mailed back to brain researchers.

Checking his e-mail one morning in the spring of 1997, Giedd was stopped short. The numbers showed that the brains he was studying were undergoing dramatic changes around puberty and early adolescence. The brain's gray matter—its outer layer—was thickening, and then dramatically thinning down, a level of change that was supposed to be largely over by kindergarten.

"Basically, I thought I was wrong. I thought the numbers were off," Giedd says.

Brain thickening generally happens when the tiny branches of brain cells bloom madly, a process neuroscientists refer to as overproduction, or exuberance. Although there's debate about this, many believe that in periods of such exuberance the brain may be highly receptive to new information, or primed to acquire new skills, particularly those related to basic survival. For years, one of the strongest-held beliefs in neuroscience was that this exuberance occurred primarily in early brain development. But Giedd had found a burst of exuberance in the teenage brain.

"There was so little information out there, and what was there said that this overproduction was over long before teen years," Giedd said. "I just kept looking at the data. Then, after about six months of looking at more brain scans, I thought, hey, this is for real."

Giedd gathered data from nearly 150 developing brains that showed the same thing. He published a paper on his findings in the highly respected scientific journal *Nature Neuroscience*, that reported that the first long-term look at a large number of teenagers had found that their brains were still growing much later than scientists had previously thought.

Giedd has now detected continued growth in a number of key areas of a teenager's cerebral cortex, including the parietal lobes, which are associated with logic and spatial reasoning, and the temporal areas, which are linked to language. And perhaps most important, he discovered complex, ongoing growth in the frontal lobes, the area right behind our foreheads, the brain's so-called policeman or chief executive, that helps us plan ahead, resist impulses, in essence be grown-up. Giedd found that the brain's frontal lobes continue to grow, peaking at puberty at about age eleven in girls and twelve in boys. And the process of change continues. After rising in volume far beyond adult levels, the gray matter in the adolescent brain then does an about-face and starts a steep trek back down. In fact, by scanning the same kids over and over, Giedd found the frontal lobes, the very area that helps make teenagers do the right thing, are one of the last areas of the brain to reach a stable grown-up state, perhaps not reaching full development and refinement until well past age twenty.

"We found a second wave of production of gray matter—more branches, more roots—and it reached its peak right around puberty," Giedd said. "Then the brain is pruned back to the essentials, you know, like one of those poems, a haiku. It's [as if] the brain says, hey, it's time to specialize."

Giedd initially looked at growth only in the gray matter of the cortex, the quarter-inch-thick outer layer or bark of the brain that's divided into areas for special functions and that, in humans, has gotten so big it's had to fold into deep wrinkles to fit inside our skulls.

Gray matter includes much of what's considered crucial in the brain: the plump cell bodies of brain cells—the neurons—and their tangle of short treelike branches, called dendrites that reach out and, acting like antennae, receive information from other neurons. It also includes many of the synapses, the spots on dendrites—in fact, they're microscopic gaps—where neurons communicate by passing chemical messages back and forth. (The brain's so-called white matter is made up of axons, the single long, stringy arms of neurons that stretch great distances across and deep into the brain and send signals to other neurons.)

Much of basic brain development is driven by genes, but many connections, some dendrite branches and their synapses, develop and thrive simply because they're used the most and grab the most neurochemical juice. That's the brain's basic modus operandi, the principle of "use it or lose it," and it means that certain life experiences—good and bad—can have an impact on the brain's essential architecture. Study a lot of Latin, for instance, and your Latin synapses carpe diem. And with that thought in mind, many neuroscientists, looking at the work of Giedd and others, have now conceded that the teenage brain is far from finished. Instead, it remains a teeming ball of possibilities, raw material waiting to be synaptically shaped. The teenage brain is not only still incredibly interesting but appears to be still wildly exuberant and receptive.

Giedd's boss at the NIH, Judith L. Rapoport, head of a group that does research in child psychiatry, says finding such extensive growth and continued activity in the adolescent brain adds new dimensions to the way we think about teenagers and their brains.

"We knew there were some subtle changes, but Jay showed powerful new curves and volume peaks at different times," she says. "I think it's safe to say that there are massive changes in synaptic reorganization during this period of adolescence. You get a leaner, meaner thinking machine.

"And if you think of the first stage as overproduction, you have to wonder," she added. "Is it just some built-in redundancy of nature? Or does it give you the ability to decide whether to be a coal miner or a violinist?"

WHAT'S NORMAL?

For years, brain scientists have tried to figure out how to measure brain growth. We have charts that tell us how a child's legs and arms are supposed to grow; we can measure the outside of heads. But we don't know how a child's or an adolescent's brain develops. What is a frontal lobe supposed to look like at age twelve? Are language areas in the temporal lobes fully grown at sixteen? Can we see how a brain is growing or shrinking as a teenager wrestles with negative integers or weeps over *Jane Eyre*?

The search is anything but idle neuroscientific curiosity. If no one knows how a brain is supposed to grow, it's impossible to tell when and how a teenaged brain goes wrong.

The new discoveries in teenage neuroscience by Giedd and others build on a series of classic studies that have attempted to find out how a normal brain grows. In one, Peter R. Huttenlocher, at the University of Chicago, actually gathered brains of children from autopsies and counted the number of tiny synapses—an indication of a brain's density of connection and development—in brains of different ages.

Looking at part of the frontal cortex, for instance, he found that synapses begin to increase rapidly before birth, reach adult levels at birth, and then continue to rise, reaching twice adult levels by age one or two. Then, after leveling off at that high plateau for several years, the synapses began a gradual slide, wiping out nearly half the connections and bringing brains back to adult levels again.

Huttenlocher's work, in certain ways, mirrored several other careful studies in this area. In the 1980s and 1990s, neuroscientists Pasko Rakic and his wife, Patricia Goldman-Rakic, at Yale, and

Jean-Pierre Bourgeois, at the Pasteur Institute in Paris, in highly scrupulous studies, counted synapses and outlined brain development in rhesus monkeys. They found a similar pattern: There was a rapid increase in synaptic density (the number of connections per neuron) before birth until, at birth, synapses rose to adult levels. By about two months, monkey synapses reached twice adults levels and remained on that high plain until about age three, when monkeys are sexually mature and synapses began to taper off.

In 1987, Harry Chugani, now at Wayne State University, found further evidence that supported these findings. Using PET (positron-emission tomography) scans that measure glucose use by the brain, he scanned twenty-nine epileptic and normal children and found that glucose levels, at birth, were about 70 percent of an adult's, and twice adult levels at age two or three. At about eight, brain use of glucose, according to his calculations, started to decline and level off, eventually falling back to close to adult levels at around age sixteen or seventeen.

All three studies confirmed the same pattern. Synapses started to form before birth, rose to adult levels at birth, increased to twice adult levels through childhood, stayed there for a period of time and then fell to adult levels.

The Rakic, Chugani, and Huttenlocher studies painted a portrait of the developing brain that showed, unequivocally, that the brain operates as much of nature does: It hedges its bets by creating more synapses than are necessary, in essence putting a lot of cards on the table and letting the players duke it out, the best and strongest brain connections winning.

None of these earlier studies, however, focused on teenagers: Rakic had studied monkeys; Chugani's had been an indirect, rather than direct, measure of brain development; and Huttenlocher's study had included very few adolescent brains.

Giedd's study filled that gap. Studying the living brains of nearly 150 teenagers, he found a similar developmental path, but with a twist. Connections in the teenage brain, he discovered, did not decline steadily from prepuberty on. Rather, he found that at a certain point around puberty, the brain undergoes a growth spurt, particularly in the area that most makes us human, the frontal lobes.

To date, Giedd is the only researcher with extensive long-term data on the growing brains of teenagers, many of whom he has scanned six times. For that reason, other neuroscientists generally don't dispute Giedd's numbers. Although he actually measured the volume, or overall size of the brain, and from that deduced growth in synapses and dendrites, the consistency of his findings is impressive.

No one, of course, is saying that more synapses make people smarter. Science doesn't know that. In fact, there are some disorders such as fragile X syndrome, the second most common form of mental retardation, where the brain seems to contain too many synapses, tangled and knotted and confused. It's also obvious that while we shed synapses in late adolescence, we don't necessarily get dumber, at least not all of us.

Yet, as neuroscientist Patricia Goldman-Rakic says, synapses are a *big* deal in the brain because they're how brain cells communicate with one another, activity crucial for any behavior. "Synaptogenesis and communication between nerve cells is essential for mediating any function," she says.

B. J. Casey of the Sackler Institute in New York, another highly respected child brain scanner, says it's now clear that the brain "is not complete at adolescence; it's still being refined."

In one of his most recent scientific papers, Huttenlocher expanded on that point, talking about what all this might mean for an average, normal teenager. "More complex executive functions of [the] prefrontal cortex such as reasoning, motivation and judgment, appear to develop gradually during childhood and adolescence, perhaps continuing during the adult years," he wrote. "These uniquely human functions appear late during development and their emergence may be aided by late persistence of exuberant synapses in [the] prefrontal cortex." For that reason, he said later, while we should expect teenagers to begin to perform higher level thinking, we also "shouldn't be surprised if high school students have trouble making decisions."

In fact, as the teenage brain is reconfigured, it remains more exposed, more easily wounded, perhaps much more susceptible to critical and long-lasting damage than most parents and educators or even most scientists had thought. Adolescence, some neuroscientists now warn, may be one of the worst times to expose a brain to drugs and alcohol or even a steady dose of violent video games.

As Giedd puts it: "If that teenage brain is still changing so much, we have to think about what kinds of experiences we want that growing brain to have."

Adolescence, agrees Harry Chugani, is a time when brains are absorbing a huge amount, but also undergoing so many alterations that "many things can go wrong." As he sees it, the teenage years rival the terrible twos as a time of general brain discombobulation.

"These two times are when children show the most astounding changes in behavior," he said. "That can't be just a coincidence."

Chapter 3

The Age of Impulse

Refashioning the Frontal Lobes

Jamie gets good grades at a competitive high school and, concerned about her future, she's careful not to do drugs or drink. Nevertheless, every now and then she gets an urge to "do something crazy."

One afternoon not that long ago, the urge struck while she was driving on a highway. A truck passed her and she suddenly thought, "Hey, why are you doing that?" Jamie pushed the gas pedal of her car to the floor and quickly accelerated to over one hundred miles per hour. She passed the truck, but she also nearly killed herself.

Did she think about that? Did she think about what would happen if she had an accident at a hundred miles an hour just because she wanted to beat a stupid truck?

"No, not really."

Another girl, Jessica, also found herself in a sticky situation, all of her own making. At her part-time job, she'd been having long telephone conversations with a man who worked in another office of the same company. One day, on a lark, Jessica, who had just turned seventeen, told the man she was twenty-one and agreed to meet him alone after work. All she knew about him was that he was forty-one years old. "I guess, looking back, he could have been a mass murderer or something," Jessica says.

But did that particular thought sink in before she lied about her age and agreed to meet him? Did she think things through, worry about the consequences?

"Not completely, no," she says now. "I just wanted to do it, so all of a sudden, I did."

These two teenage girls, studious, nice, smart—and at times, out of the blue, ever-so-slightly wild and crazy—are hardly alone among the ranks of the world's adolescents. When Julia from Long Island, was thirteen, she would, on a whim, regularly grab on to the back of an ice cream truck to get a fast ride on her Rollerblades. Ian skateboarded down city streets so fast that, he says, often "I fell on my head." Lisa, fourteen, on a dare, agreed at eleven P.M. one night to walk through a scary part of town because she "just didn't think about anything bad happening and just wanted to do it." Late one winter afternoon, Jack and a group of his fifteen-year-old friends, after watching an adventure show on television, decided on an impulse to take a shopping cart from a supermarket to a nearby park, scoop a downhill track in the snow, and then take turns careening down the hill until they flipped over on the ice. Jack wound up with a whopping headache and a concussion.

"He just didn't think," Jack's mother said. "I told him, if that's the kind of judgment you use when you're sober, don't you ever, ever drink."

PLENTY of teenagers are frightened, and a few are more thoughtful and careful, by far, than their parents. But most, every now and then, get an irresistible urge, as Jessica said, to "do something crazy," to be impulsive. It's one of the world's stereotypes about teenagers that just happens to be true. And it's hardly new.

A woman in her fifties, who was one of the first females admitted to Yale, remembers when, as captain of her high school field hockey team, she decided, along with some friends, to paint "Let the Good Times Roll" in five-foot letters on the side of a school building, thinking that they "wouldn't get caught." Jeff, fifty, a teacher and business consultant in Santa Cruz, California, has a picture-clear memory of the afternoon he spent at the police station after he and a friend, armed with their teenage brains and a long-range fire extinguisher they had somehow gotten their hands on, shot streams of water at a group of young Boy Scouts picnicking in a park. Gerard, a writer now in his mid-forties, can still recite moment by moment the events of a hot June day when, after a softball game in Central Park, he and a group of his fifteen-year-old friends decided it would be a good idea to jump into the large fountain in front of the Metropolitan Museum of Art, yelling and laughing, their soggy sweatpants dragging behind.

"Was it impulsive, I'd say yes, very. I mean, it's not so much that no one thought it through, no one thought at all," says Gerard. "But isn't that what teenagers are all about, impulses, twitches?"

Now with teenagers of his own, Gerard says some of the impulsiveness he sees in his own children has developed a distinctly modern flavor. His sixteen-year-old daughter recently started spending hours sending instant computer messages to her friends. She ignored time, she forgot her schoolwork, she couldn't stop.

"We had to limit it; she was almost addicted. She's a good student but her grades went down and she didn't think of that possible consequence at all," Gerard said. "She was only thinking of the tingle of the moment."

AT THE Institute of Child Development at the University of Minnesota in downtown Minneapolis, Chuck Nelson works as a neuroscientist trying, by scanning the brains of normal kids, and by studying the effects of deprivation on Romanian orphans, to understand the intricate, complex development of the human brain. As the father of a fifteen-year-old boy, he also spends a fair amount of time trying to understand that complex creature as well. And Nelson is only too familiar with the impulses, the twitches, the attraction to the tingle of the moment.

There was the time, for instance, when his son, a good kid, decided to paint graffiti on a building downtown. There was a long period when his son would "blow up" angrily at the slightest remark by his parents at the dinner table.

Recently, though, Nelson's son had shown signs of change. When he came home fifteen minutes late one night recently, instead of exploding in response to his father's anger, Nelson's son had calmly apologized, promising it would not happen again. An ice hockey goalie, he now sat for hours watching other goalies, taking mental notes, planning future moves on the ice. There were still things he didn't get, like "the idea that if you mow the grass right away, you can see your friends faster." But by and large, Nelson had seen his son take great leaps toward growing up.

On a recent trip to Montreal to talk with neuroscientists there, I'd taken my own daughter, Hayley, shopping. The exchange rate was good, the clothes were French and chic, and I'd told Hayley, then sixteen, she could buy school clothes.

"No, not now," she told me, a somber look on her face. "I think I'll wait. I'm planning on buying things in California when we get there."

What? Wait? This was the girl who never slowed down enough even to try clothes on before buying them. A plan? Wait? What was this?

When I mentioned the shopping incident to Chuck Nelson, he just nodded his head. To him, these stories—his son's spray painting and blowing up, as well as his newfound calm, my daughter's newly acquired ability to wait and plan—pointed to one place in the teenage brain: the frontal lobes.

The frontal lobes, one of the key areas Giedd's research found still under construction in adolescence, are the part of the brain that helps us resist impulses, wait before spending all our money on clothes, stop before we yell regrettable things at well-meaning fathers and mothers.

"Basically, this is the part," Nelson said, "that tells you to count to ten before you call your mother old and stupid."

The Prefrontal Cortex

When scientists talk of the frontal lobes of the brain, they're most often referring to the prefrontal cortex, the section of the brain located directly behind the forehead. Human brains are little wads of evolutionary history. Some parts, such as the brain stem and parts of the inner limbic system—areas that regulate emotions and gut reactions, the "I'd like to eat you now" instincts—we share with your basic crocodile.

But the prefrontal cortex is something else again. Neuroscience textbooks are often illustrated with stark diagrams that show how the prefrontal cortex has grown disproportionately huge in human beings. First, there's usually a drawing of a rat's brain with its tiny smooth prefrontal cortex, followed by a cat's brain with a bigger prefrontal cortex, then the monkey with its bigger-still brain and prefrontal cortex. And then, by way of finale, the human brain, outsized, folded in on itself with its puffed-out prefrontal cortex. By some estimates, the human prefrontal cortex, over the course of evolutionary history, has increased a whopping 29 percent, while the same region in our closest relative, the chimpanzee, has grown perhaps 17 percent and that of a cat only 3 percent.

Development of a baby's brain in the womb roughly follows a step-by-step path. The brain stem and parts of the limbic system develop early, only later followed by the wrinkled prefrontal cortex and other more sophisticated sections, in rudimentary forms. (Interestingly, parts of the late-developing frontal cortex are also often the first to disintegrate in degenerative diseases such as Alzheimer's. Scientists speculate that that might be a price the area has to pay for its sophistication. It may simply get worn out faster than other parts of the brain due to its life of plasticity, its unparalleled knack for interacting with and adapting to the environment.)

The frontal lobes, in other words—the same parts that scientists have now found to be undergoing so much change at adolescence—are a big deal in the human brain. They also have, to my mind, some of the most intriguing stories. Neurologist Oliver Sacks, author of *The Man Who Mistook His Wife for a Hat* and other books of neurological oddities, once told the story of a man with damage to his frontal lobes who became, through a distinct lack of impulse control, a compulsive "toaster." "He would rise to his feet when dining in restaurants," Sacks wrote in the journal *Neurology*, "clear his throat loudly to command attention and propose a toast to Her Majesty the Queen. The diners around him would be surprised, but rise to their feet and obediently lift their glasses. A minute or two later, the performance would be repeated, this toast, perhaps to the Lord Mayor of London. Other toasts would follow at frequent intervals, until his embarrassed family had to take him from the restaurant. The patient . . . seemed to enjoy his toasting but (while of considerable intelligence) showed no insight into it."

On Chuck Nelson's wall at his office at the University of Minnesota is a picture of a white skull with a red pole sticking out of the forehead, the head of Phineas Gage, the subject of one of the better-known stories in brain science lore.

Phineas had been an extremely capable and likable fellow until, while working as foreman on a railroad in Vermont in 1848, an explosion sent a thirteen-pound iron tamping rod through the front of his skull. After the accident, Phineas could function fine physically and converse. But he was a changed man. He lied, he stole, he cursed. He became impulsive and could not, for the life of him, plan ahead.

It was Phineas who started scientists speculating in the nineteenth century that the front part of the brain was somehow quite important in a whole range of humanlike behavior, particularly the ability to inhibit impulses and plan. More than a hundred years later, Antonio Damasio and his wife, Hanna, two of the country's leading neuroscientists at the University of Iowa, examined Phineas's skull, preserved at Harvard, and, using modern imaging techniques, pinpointed the site of the damage. The rod had, indeed, gone through a section of the prefrontal cortex.

IN RECENT YEARS, neuroscience has been engaged in an all-out effort to zero in on exactly what this important brain part does and how it does it. In the 1970s and 1980s, neuroscientists Patricia Goldman-Rakic at Yale and Adele Diamond, now at the Eunice Kennedy Shriver Center in Massachusetts, did a series of elegant experiments that showed when and how the prefrontal cortex kicks into action. The tests they did—Goldman-Rakic on rhesus monkeys and Diamond on human infants—are called delayed response tests, or A not B tests, designed to find out how long a subject can retain information that is no longer in view. To do this, a monkey or child must hold a representation of the picture in mind and inhibit the impulse to let other irrelevant information take its place.

The A not B task has been linked to a function of the prefrontal cortex, more precisely the dorsolateral prefrontal cortex, that neuroscientists often refer to as working memory—the ability to keep a seven-digit phone number in your mind just long enough to dial the number, for instance. It is believed that working memory, often referred to as the brain's blackboard or Post-it note, is linked to impulse control. As Goldman-Rakic puts it: "If you are not able to direct responses by mental representations, you also are not able to withhold reflexive responses to irrelevant or salient stimuli." In other words, if you can't inhibit your brain from responding to every urgent e-mail from your friends, you'll forget your homework again.

In the Goldman-Rakic and Diamond tests, monkeys and humans did versions of similar tests. First they watched as a bit of food or toy was hidden in one of two or three receptacles, and then these receptacles were blocked from their view. After a few seconds, the receptacles were brought back and the animals and children had to remember which one contained the hidden food or toy.

Young monkeys and young infants largely flunked. But the researchers found that as they got a bit older, they could keep information about the item's location in their heads for longer and longer periods. Monkeys got good at the task at around four months and human infants started to improve at an equivalent age of about seven months. By a year, the human babies could remember where the food was hidden for ten seconds.

As it turned out, the monkeys and babies had acquired a hugely important basic skill as, according to timetables established in earlier studies, synapses in their brains' prefrontal cortex began to flourish. And later studies showed that the precision of this skill improves up and through late adolescence along a path that mirrors the delayed refinement of the prefrontal cortex.

From this experiment and others that track development of functions—for example, vision—many scientists have concluded that a flourishing or exuberance of synapses in the prefrontal cortex—an overproduction similar to what Giedd found in adolescent brains—may be linked to establishing and refining crucial brain functions.

As Goldman-Rakic explains it, the brain must be at a stage of readiness, in terms of densities of synapses, before it acquires certain basic skills. "The brain seems to have to get to a certain level of circuitry, to be fully wired, before certain behaviors emerge."

Through the years, neuroscientists have come up with plenty of additional tests to explore what the prefrontal cortex does and how it develops as children age. One is the go-no-go test in which the subject, whether monkey or child, is first given a choice of which button to push, blue or yellow. They quickly learn that if they pick blue, they get a reward. So, they keep picking blue. Then the rules change, and the reward comes only if yellow is pushed.

Young monkeys, young children, and those with severe damage to their prefrontal cortices, generally cannot respond quickly to such changes. They are at the mercy of their natural impulse to keep picking blue. As children age—again along a track that follows, to varying degrees, the continued and prolonged development of their prefrontal cortex—they get considerably better at adapting to changes in rules. Increasingly—improving well past age twelve—they can control their instinctive impulse to keep picking blue, and instead, stop, think, and, when appropriate, pick yellow.

As John Mazziotta the neurologist at UCLA said, "People don't realize that the brain is really an inhibition machine." To explain how all this works, Mazziotta talked a bit about imitation. One crucial way humans and other creatures learn, Mazziotta said, is through the art of imitation. Humans are programmed to imitate.

"Watch me," he said as he sat behind his large desk at his university office. He pretended to raise a cup of coffee to his mouth. "Do you know that your brain was essentially doing the same action as you watched me? We're creatures of imitation, that's how we learn. But the brain has to be able to inhibit inappropriate actions, including imitation. That's how it works."

Mazziotta pulled out a neurology textbook with pictures of a woman kneeling and praying next to a man who was also kneeling and praying. The woman, Mazziotta explained, had suffered brain damage and could no longer inhibit certain actions. She had not the slightest interest in kneeling and praying at that moment, but she could not stop herself from doing what brains want to do, imitate the action they see, like a monkey behind the glass at a zoo making faces back at you.

Another thing to remember, Mazziotta said, is that many of the brain's systems are running all the time. "Think of an airplane," said Mazziotta. "Most people think that when it lands it has its engines on low and it's just floating in. But that's not always so; in landing, an airplane often has to be at full throttle in case it has to react quickly if something happens." The brain, too, he says, is set up to be whirring all the time. Even when we think of it as resting, its neurons are often firing at a low level, ready and waiting, so it can react in time before, for instance, it's eaten by a bigger, quicker brain.

The brain is working constantly, and one of the tasks it works at is to inhibit itself from a variety of actions. It is striving to resist the urge to raise the coffee cup like the guy across the table, and striving *not* to do a number of things that might not be in its best interest. As the brain develops—in children and, science is now learning, in teenagers—it is this very inhibition machinery that is being fine-tuned.

"Development," says Mazziotta, "is progressive inhibition."

Or, as another neuroscientist said, in explaining the importance of the frontal lobes in inhibiting inappropriate actions: "When this front part of the brain, this prefrontal cortex, declines in old age, what do we get? Old men grabbing nurses in nursing homes, that's what."

THE MYTH OF THE ADULT CHILD

And what can we expect of adolescents if that inhibition machinery, the prefrontal cortex, is not yet fully tuned? If it is still being remodeled and reorganized as suggested by Jay Giedd's study, as well as the studies of other scientists, could that help explain some of the well-known impulsive, distinctly uninhibited behavior of the teenager?

Or, looked at in another way, could a more mature and finely tuned prefrontal cortex explain a newfound ability to resist that need to buy the first clothes you see or impulsively blow up in anger at a parent? Can we now say that such fundamental shifts in behavior are likely linked to actual developmental and structural changes in the architecture of the teenage brain?

"No question," says Chuck Nelson. "A lot of teenagers just don't see consequences of actions. They don't think ahead. They don't see that getting good grades today, for instance, makes a big difference to the person they will be later on. When they get older, they start to get that. And I think it has to do with development of the brain, particularly the prefrontal cortex, the part that controls working memory, inhibition, impulse control."

But if impulse control can be linked to brain development, why is it teenagers often seem, in some areas, even more muddled and clueless than eight-year-olds? After all, an eight-year-old's prefrontal cortex should be far less mature than a thirteen-year-old's.

One reason, Nelson suggests, is that, as children get older, they call on their still-unfinished prefrontal cortex more. The world gets more complex, school gets harder, social relationships get more obtuse. Adolescents have bigger passions, too. "They need to be independent from their parents; they want to be adults and they're exposed to a semiadult culture. But they don't have the prefrontal cortex to regulate those adult behaviors; they drink and they drive without seatbelts, all of that."

Or as Giedd puts it: "They have the passion and the strength but no brakes and they may not get good brakes until they are twenty-five."

It's easy to be overly optimistic about a teenager's braking system. We look at kids who are taller than we are—my thirteen-year-old daughter has size-ten feet—and we think they ought to instantly act like us. This is what some have dubbed the myth of the adult child, an idea that many believe has become pervasive today, particularly, as one sociologist I spoke to said, "baby-boomer parents try to be friends" with their children. Mark Howard, head counselor at a teenage residential drug rehabilitation center run by Phoenix House on Long Island, points to this as one of the major problems with many of the kids in trouble he sees.

"Kids today are getting mixed messages," he says. "Parents want to be friends with them; they don't set good boundaries. I don't think that a fourteen-year-old has what I would call consequential thinking. They just don't. But parents go back and forth about that and don't seem to understand. They will crack down and tell a kid that he can't watch some TV show, but then they will leave that same kid alone in a $250,000 house and think they are responsible enough to make the right choices in that situation. And guess what, most of them aren't ready for that. They don't think things through. And sometimes they get into a lot of trouble."

One mother I know in New Jersey found this out the hard way. When her daughter was a sophomore in high school, she somehow got hooked up with a "group of kids who wanted to break the rules." She "liked the excitement" of being with the bad kids, started cutting school and, one night, in a house where the parents weren't home, she had sex, at age fourteen, and got caught.

"We were at our wits' end, we didn't know what to do," the mother said. "So I asked a wise friend who said that sometimes you simply have to be the parent. So we told her that unless she changed her friends, she had to change her school. I think she wanted someone, some adult, to tell her to just stop. And it worked. She joined the lacrosse team and the high school swim team where she was elected captain. Her grades went back up and she was back to herself again."

Peter Jensen, a former head of child and adolescent research at the National Institutes of Mental Health (NIMH), now director of the Center for the Advancement of Children's Mental Health and a professor at Columbia University, a practicing child psychiatrist, and, perhaps most important, the father of five who's made his share of mistakes along the way, says he has learned a lesson from dealing with adolescents in his clinical practice and in his own home. Sometimes, he says, parents need to act as though they are their teenagers' "frontal cortex."

"We like to think that maturation is based a lot on experience, but even in adolescence we also have to recognize that learning may not count so much until the underlying brain structures are in place," Jensen says.

While waiting for those structures to develop—and perhaps helping them get set up right in the first place—Jensen says parents of teenagers often have to "walk a tightrope." On the one hand, they have to respect and encourage their teenagers' need for autonomy because, in adolescence, "that's where the action is." But sometimes they also need to step in, offer a road map, and help those teenagers point their size ten feet down the right path.

To do that effectively, he says, parents might take tips from some of the ways that psychiatrists, through the years, have found to deal with teenagers. Parents, says Jensen, might try acting a bit like the psychiatrist played by Judd Hirsch in the movie *Ordinary People*, talking through possibilities and options. They have to function like a surrogate set of frontal lobes, an "auxiliary problem solver."

"With little kids you can tell them what the best thing to do is and then offer them a reward—tell them that their mommies will be very happy if they do this or that," says Jensen. "But with teenagers that's not often a productive approach. If you just flat out tell a teenager what to do, you can lose that kid. You have to cut them some slack, but you can't just leave them there, you also have to help them figure out things themselves. You can say, 'What do you think the consequences will be if you act a certain way?' for instance, or 'What will happen if you are rejected by your peers if you reject drugs?' "

In many parts of the world, of course, kids are considered adults at ages as young as twelve. But, as Jensen points out, those societies often are "more structured and ordered" than our own. In modern Western culture, life is considerably more complex. As Jensen says, there are "so many more stimuli and so many more opportunities; so many more ways to go wrong," that it's "lucky we have such a protracted period to protect" adolescents as they are trying to get it right.

"There's an old psychological notion that says that things like college or the Peace Corps or even the military are good simply because they allow for a psychological moratorium," Jensen says. That theory, he says, was referring to "psychological crystallization." But "when you think of it, it could also allow for brain development." It allows that teenage brain to try out its newfound abilities to plan ahead, to think things through, to resist impulses—to, as Jensen says, "learn how to do it; to practice."

An average teenager gains fifty pounds and grows a foot in the space of four to five years. At the end of that growth spurt, that teenager may outwardly look like a mountain of maturity to us. But it's an illusion.

As Chuck Nelson in Minneapolis puts it, "You don't suddenly get perfect synapses. It takes a while to get the right connections working smoothly. This may be the speed bump of adolescence."

Chapter 4

Altered States

How Experience Changes the Very Structure of the Brain

Marian Diamond, one of the grande dames of neuroscience, is known for her pioneering work on how experience molds brains.

She's also a delightful piece of work herself. When we met at her office at the University of California at Berkeley, her white hair was piled high on her head, she wore a bright purple mohair sweater, deep blue eye shadow, red fingernail polish on long nails—and instantly offered her guest chocolate-covered peanut clusters because, she insisted, "they're very good for the brain."

Diamond's work, for the most part, has been with rats. She loves rats. Holding up her hand to her neck as if cuddling a big fat rat, she told me: "They have a bad reputation, but they don't deserve it, you know. They're so soft and nice and very smart."

As a grandmother, Diamond knows that teenagers, the human variety, also have a bad reputation. But as far as she's concerned, they don't deserve it either. Teenagers can be annoying, perplexing, peculiar even. But, like rats, she says, they're also very smart.

And one of the major ways they get smart, she believes, is from the specific, individual connections built in their brains by the specific, individual experiences they have.

"What does it mean that the brain is still growing at adolescence?" she said when I asked her what she thought of the new work on the teenage brain. "Why, dear, it means everything. Everything, do you understand? The brain is everything."

THE idea of experience and brains took root a number of years ago in Canada, largely by accident. In 1949, Donald Hebb, one of the early neuroscientists, was working at the Montreal Neurological Institute when he decided to bring his children's pet rats to work. Hebb, apparently as fond of rats as Diamond, not only let his kids keep rats as pets, but the rats ran all over his house as well. On a whim, Hebb put the home rats in mazes he used for his lab rats and, to his surprise, the pet rats ran the mazes much faster. Could it be, he thought, that careening around the varied and unpredictable environment of a human house had somehow made the home rats smarter, had made their tiny brains grow?

Later, at UC Berkeley, as a young researcher, Diamond joined the team of Mark Rosenzweig, David Krech, and Edward Bennett, who had decided to test that idea. The scientists put some rats in a cage with toys—little ladders and wheels—as well as other rats for company. Other sets of rats were plunked in nearly empty cages, without toys or friends. After a few months, the researchers dissected the brains of all the rats, slice by slice, and counted what they'd found, neuron by neuron.

And what they found was a little revolution in neuroscience. The rats with toys, the ones that lived in a more complex environment, had a thicker cortex; they had many more glial cells, the cells that help nourish neurons in the brain. Their brains also had wider spaces between neurons. Though the researchers did not count synapses in their initial study, they hypothesized that the wider spaces meant there were more dendrite branches and synapses in the brains of the toy-playing rats.

In other words, the complex environment had made the rats' brains grow and change; the toy-playing rats were also quicker at mazes. In 1964, the team published a paper saying that experience could change the fundamental structure of the brain. With the work of Diamond and others, neuroscience has continued to push this connection between experience and brain growth. One of the scientists who has done the most work is Bill Greenough at the University of Illinois.

Greenough has shown, in various studies, that complex environments make rat brain synapses and dendrites increase, regardless of age, although changes in younger brains are generally more impressive. He's also found that rats with more synapses do mazes faster.

The studies have been done mostly on rats, and some scientists doubt such consistent results could be found in primates or humans. But the findings, nevertheless, have overturned old ideas that experience does not alter the brain in any fundamental structural way—the long-held view that we're born with all the synapses we'll ever have and all they do, as we age, is get stronger. And any links between experience and brain development are likely to apply to teenagers as well.

Greenough said that it had been clear to him, while watching his own daughter grow up, that something big was going on in terms of brain development in the teenage years. "Think about the tasks of an adolescent and a child," he said. "They're developing social coordination and language capacity and cognitive function. It's certainly not surprising that the brain region most involved in those things, the prefrontal region, is required to grow the most and store the greatest amount of information during that time." In the teenage brain, Greenough said, there are "cognitive leaps," as well as a gradual slowing down of other abilities, such as the ability to learn a foreign language without an accent.

"After adolescence, it's rare to find a person who can learn and speak a language that is accent-free. There's something fundamental about how the brain becomes transformed through that period," he said.

GENES OR BLUE JEANS

Greenough has developed a system for determining what kind of experience affects brain synapses at what point in time. Some synapse change seems to be pushed by genes, what he calls "experience-expectant." This is the kind of change that's supposed to occur, that's expected to occur, in every member of a species in a normal environment, such as development of vision, hearing, and some parts of language. Exposed to normal sights and sounds, a normal brain sculpts its exuberant excess of synapses into appropriately connected networks so it can respond to such things as the basic sounds of its own mating song or language. If that brain is not exposed to those fundamental experiences—a mother's voice, the outline of a tree—it can develop abnormally, perhaps being unable to sing the right song to attract the right mate at the right time, for instance.

But there's another kind of change Greenough labels "experience-dependent." This is synapse growth that depends more on the kind of experiences an individual has—does that teenager live in Kosovo or Kansas?—the kind of growth that makes a brain unique and able to adapt to wherever it finds itself. Like many neuroscientists today, Greenough is continually impressed with the brain's "plasticity," its ability to adapt and change. His own recent work suggests that even things like tiny capillaries in the brain appear to "increase with demand." Brain measurements of professional violin players have shown that areas of the cortex devoted to the fingers on their left hands (the fingers that fly across the frets) are considerably larger than corresponding areas in the brains of nonviolin players, and in younger players, the areas appeared to grow more easily than in older players. Other studies have shown that some stroke victims, after highly intense therapy, can expand specific brain areas and regain use of paralyzed limbs.

PLASTICITY—PRO AND CON

Clearly, the brain needs to be, in some respects, unalterable. Otherwise, we would not retain a sense of our selves or even remember where the refrigerator was. But the human brain must also remain changeable, plastic, so we can survive in changing circumstances. This is as true, and perhaps even truer, with teenagers as with anyone.

Of course, as Chuck Nelson in Minnesota points out, not all environments are healthy in terms of brain development. The environment in which our brains, teenage or otherwise, find themselves can also be toxic. And that can produce correspondently toxic results in the brain. As Nelson puts it: "Plasticity cuts both ways."

Nelson has watched firsthand the experiences of Romanian orphans, many of whom were raised in horrible conditions, stuck in cribs and never held, never hugged. Later, adopted by foster parents, many of the orphans as babies did not cry, and years later many remained disruptive and behind in school. While Nelson and others have now shown that some of those early bad experiences can be overcome with exposure to healthier environments, it's also clear that the orphans who are rescued earlier fare much better.

"These are the powerful effects of experience," Nelson said. "You have an exuberance of synapses that possibly lets you take advantage of what's going on in your environment. But what's going on in that environment can be toxic, too."

A recent study by Harry Chugani, who scanned the brains of some of the Romanian orphans in foster homes in this country, found that their brains did seem to be different. While the impact of early deprivation in animals is well-documented, there are few detailed studies of what happens to the brains of maltreated children as they age. In this PET scan study of ten orphans, all around nine years old, many of the children's brains, Chugani said, were found to be "metabolically less active" in parts of the inner brain limbic area linked to recognition of faces and emotions, two crucial components of bonding and attachment.

The study was small and included only severely deprived orphans, but it nevertheless highlighted a possible link between toxic environments and human brain development. The study's conclusion? "We suggest that chronic stress endured in the Romanian orphanages during infancy in these children resulted in altered development of these limbic structures and that altered functional connections in these circuits may represent the mechanism underlying persistent behavioral disturbances in the Romanian orphans." In other words, as Chugani told me, "There probably is a sensitive period when those structures have to be used to develop correctly."

There is little doubt in the minds of Harry Chugani and Chuck Nelson that experiences also alter the developing synapse-rich brains of adolescents. But they, along with others working in this area, are gun-shy about too blatantly linking specific brain growth to specific human behavior.

"The thing is," Nelson said, "we know experience matters, but we just don't know what nature of experience matters, what's best for the brain." As a parent, though, Nelson can't help but have his own thoughts on the subject, based on hours of gathering his own anecdotal evidence.

"Think just about this, for instance," he told me. "You have kids. Tell me, do you think you influence them more by lecturing them or by those spontaneous talks you have when they're in the backseat of the car?"

THE CEREBELLUM

The dispute over nature and nurture is an old one, of course. How much of a developing teenage brain can you change with experience? And how? Is it all predestined by a tiny, twisted coil of DNA or can we, through casual backseat conversations, toy-filled rooms, good luck or bad, alter the destiny of synapses and the behavior of a human being, let alone a teenager?

Intrigued by this seemingly intractable question, the NIH's Jay Giedd has started a new project that is specifically aimed at sorting out the differing influences of nature and nurture with teenagers and their brains. He is engaged in a longitudinal study of twins, trying to determine how two brains, largely identical at birth, are altered as they bump and bumble through their individual experiences in life.

In one groundbreaking finding from that ongoing study, Giedd, after examining the brains of hundreds of identical and fraternal twins, uncovered hints of which parts of the developing brain seem to be the most amenable to change from the outside. To his own surprise, his study found that the part of the brain that was the "least heritable," the part that seemed to change the most between identical twins as they interacted with their environment, was the cerebellum, the rather nondescript lump of brain tissue at the top of the neck.

For many years, neuroscience didn't pay much attention to the cerebellum. It was clearly connected to certain types of movements, but little else was known. Lately, though, there's been a bit of rethinking about the cerebellum and its role. It's been found, for instance, that kids with Asperger's syndrome, a type of mild autism that is often accompanied by social awkwardness and aloofness, have differences in the metabolism of their cerebellums. That and other findings, particularly with patients who have cerebellar injuries, suggest that it might be much more important than previously thought in a wide range of behaviors, including recognizing social cues, even getting jokes.

Giedd also found that the cerebellum continues to change throughout adolescence. In fact, it's apparently the last structure of the brain to develop, in many cases not finishing its remodeling project—its growth and pruning of brain connections—until after the frontal lobes.

"One of the tasks of a teenager is to be fluid with social interaction and to get subtle clues and even know when someone is joking," said Giedd. "All those things scream cerebellum. And now, it seems, it's also the latest maturing of all—it keeps getting bigger and better throughout the teenage years and into the early twenties. That's simply amazing when you think of it."

Even with that knowledge, of course, it's far from certain how much of an influence we can have on the developing brain.

"We may find out that all we can do is tinker around the edges, that the brain is largely genetically programmed—a finely tuned Ferrari—and it will take real sci-fi answers to help—genetic engineering and all that," Giedd said. "But we might find out that there are things we can do to improve things. My guess is that, if that is so, it's going to turn out to be something we already know about. We know that early development is important, that you need good parents—we know that.

"And we could find out that the way to make a better brain is not through four hours of homework," continued Giedd, who says that, after what he has learned about how the brain grows, he often lets his own four kids decide for themselves what they want to do if they have free time together, rather than dictating what they will do.

"What if we find out that, in the end, what the brain wants is play, that's certainly possible," he said. "What if the brain grows best when it's allowed to play?"

A "Critical" Time

Tucked at the end of one of Giedd's papers on the teenage brain lies a question: If the teenage brain growth includes overproduction, the exuberance of brain branches and synapses, as well as a large-scale cutting back of those brain branches, it may mean that adolescence is what neuroscientists call a "critical" period, a "stage of development," he wrote, "when the environment or activities of the teenager may guide" the pattern of brain growth during adolescence.

Critical or sensitive periods are small windows of time during which certain experiences must occur for proper development. Science has pinned down a number of very specific critical periods. One is imprinting, a phenomenon that occurs in some animal species, for example, a gosling following around the first big object it sees during a certain period after birth, thinking it is its mother.

In 1981, David Hubel and Torsten Wiesel won the Nobel Prize for showing that eyes had to be exposed to certain visual experiences (looking at patterns of everyday objects, for instance) during a small developmental window soon after birth or visual fibers in the brain's cortex would not connect properly. The researchers sewed shut one of the eyes of newborn cats and monkeys and, after a few months, the animals were blind in that eye, even though the eye itself was fine, something that would not happen with comparable experiences in older animals. This relationship is well established, and is the reason pediatricians treat eye problems in children, such as strabismus, in which the eyes do not properly focus together, as early as possible. If the eyes are not wired correctly during their critical period of brain development, they will not work.

No one knows exactly why brains have evolved this way, though one good guess is that there simply aren't, and could not be, enough genes to determine exactly what each of the billions of neurons will do in various environments. Genes provide a rough outline for the brain that then uses its specific environment to fine-tune its development, perhaps getting good at some things and letting other possibilities or options fall aside.

Another classic example of a critical period in brain circuitry involves language. A number of birds must hear their species mating call at a certain point after birth or they won't learn it and successfully mate. Closer to home, studies have found that while newborn human babies can discern sounds in any language, their brains quickly become wired to respond most easily to the sounds of their own language. That's why it can be hard for a person raised to speak Japanese, which has no separate l and r sounds, to hear and speak those letters in English words. It also becomes gradually harder to learn a foreign language without an accent after puberty, presumably because parts of the brain have become "hard-wired" to speak and hear one's native tongue, a phenomenon often referred to as the Kissinger effect, because former secretary of state Henry Kissinger, who arrived in this country at age twelve, spoke English with a heavy German accent, while his younger brother, who arrived in this country at age ten, was accent-free.

The whole idea of critical periods with born-to-be-adaptable humans remains controversial, and some scientists are hesitant to talk about them at all. It's obvious that the edges of some critical periods can be pushed if motivations are high enough—through months of intensive practice, for instance. The critical periods that exist in humans seem to relate mostly to basic survival behaviors, not the ability to play lacrosse, for instance. But, in any case, they do seem to be linked to times when the brain has an overabundance, an exuberance, of synapses.

And so, if there's synaptic exuberance around the time of puberty, as Giedd suggests, how does it affect teenage development and behavior? Is adolescence, in fact, a critical brain period? Is there some basic survival skill the teenage brain primes itself for? Is there some task that teenagers are supposed to master, or experiences they should have in their rather wild window of opportunity we call adolescence?

Most neuroscientists will say they do not yet know the answer. If he has to guess, however, Giedd says that any classical critical brain period in adolescence might very well be connected to one of the most classically critical of human activities, one that clearly blossoms in significance in the teenage years—mating.

"The point, particularly at that age, is to enhance your chance of reproducing, right? If you think that you have to study a lot to get a good job in order to get a good mate, or you have to go to the mall to get attractive to get a mate, all that could be part of it. The chore of adolescence is to figure out complicated social hierarchies, and obviously culture has a lot to do with that. But a lot of the activity—looking right, listening to the right music—could be seen through that filter of trying to be popular." Looked at this way, many teenaged behaviors may have their evolutionary roots in the drive to mate.

Chuck Nelson says that adolescence may well be, if not a classic critical period, at least a highly sensitive window of opportunity. Greenough agrees, but says rather than thinking of all of adolescence as one broad critical period, it might be that different sensitive periods exist in the teenage brain for a range of "subsystems," such as budding sexuality. "The hormones could bring out certain behavior, like play, that are important for sexuality," said Greenough.

Elizabeth Bates, a cognitive neuroscientist at the University of California at San Diego, believes many critical periods pertaining to language are overrated, that there are gray areas when change can occur if people are motivated enough. Nonetheless, she, too, argues that much of the behavior we consider bizarre in teenagers, even the "muddle" they often seem to be in, may, in fact, be traced back to crucial stages of development in the adolescent brain.

Indeed, she suspects that some teenage behavior comes from what she calls the "whiplash" that the brain experiences as it goes through what she calls a "massive overhauling" to respond to a "whole new environment and radical shift in priorities."

"I would be very surprised if the brain was not seriously rearranging itself at adolescence; all of a sudden they have to pay attention to this, not that," said Bates, whose own daughter successfully made her way through the adolescent rite of passage and moved on to Stanford University.

"Think of it," she told me. "The brain is getting ready for reproduction for goodness sake."

TO MANY, this idea of the brain responding to experiences, connecting up synapses as we go, is the take-home message about brains overall. Marian Diamond at UC Berkeley believes it so strongly that, even in her white-haired stage, she continues to expose her own brain to as many diverse experiences as she can.

That's why she has not retired, why she swims and walks at least a half hour each day, why she's obsessed lately with planning an "enriched" orphanage in Cambodia, where the children, who have so many needs, will get an environment early on that is, like the rats with the toys in their cages, set up to be a bit better, more enriched, for the sake of their brains.

After what she has learned in her own lab—and in raising her own children—Diamond never misses a chance to enrich her own brain and is constantly on the lookout for opportunities to sneak some of that enrichment of experience into the brains of others. After we'd finished talking in her Berkeley office one summer morning not long ago, Diamond opened her door and found my two teenagers sitting on the floor in the hallway.

"Hey kids," she said instantly, "have you ever seen a real brain?"

A minute later, we were in a lab next door, and she was taking the top off an old-fashioned, blue-flowered hatbox on the lab table.

There, inside a Tupperware bowl, was a human brain. Both kids gasped. Diamond beamed. She pulled on white plastic gloves, gave us each our own gloves, and then carefully lifted the brain out of its Tupperware home, handing it to us, one by one.

I could not help thinking, as I stared at that brain in my hands, of all the hours I had spent in English literature classrooms a stone's throw away from this lab, going line by lovely line over the poems of John Donne and John Keats. Which, I wondered, would end up the better way to understand human passions and behavior: exploring the intricacies of this mass of dendritic Jell-O I now held in my hand, or through the succulent stanzas of poetry?

Emily Dickinson herself once wrestled with this question, tipping her hat to the brain, albeit in a poem:

The brain is wider than the Sky
For put them side by side
The one the other will contain
With ease and You beside.

The brain in the hatbox, brown and pickled in formaldehyde, no longer contained any sky, of course, but it still left an impression that rivaled the most profound of poems. Each girl took a turn holding the brain, fingering its folds, peering into its crevices.

"Ohmigodomigod," said one. "You mean there's a whole personality in there?" asked the other.

Diamond patiently answered all questions, adding advice of her own. To develop properly, she told us, the brain must have certain experiences: good diet, exercise for good blood flow, challenges, and love.

"You know, I say that part about love in my lectures and the men all laugh. They are scientists and they know it's true, but they won't say it," she said, as she carefully tucked the brain back in its Tupperware bowl and closed the lid on the flowered hatbox. "Then, after the lectures, you know what those men want? They all want a hug."

Chapter 5

Making Connections

Growing and Pruning Toward Maturity

Dr. Francine Benes, dressed elegantly in a beige silk suit, was peering down at a flat brain slice neatly preserved in a glass slide. The slice—it looked like a small piece of cauliflower—was from the brain of a seventeen-year-old boy who'd been killed in a car accident.

Within seconds, Benes found what she was looking for. We were in a science lab in an old brick building at McLean, a psychiatric hospital and research center outside Boston. She directed my gaze toward a small jagged line in the middle of the slice, lit up on a screening table.

"You see," she said, suddenly excited. "Here! Here's where we found it!" Benes, a McLean researcher and a professor of psychiatry and neurology at Harvard Medical School, is another of the grande dames of neuroscience. As a young woman, she worked at a mental hospital for schizophrenics. The experience made a lasting impression on Benes, who devoted her career to understanding how the brain grows, in particular how it grows wrong in schizophrenia. And because schizophrenia most often begins in adolescence, her work has naturally turned to teenage brains.

On a visit to the Yakovlev brain bank in Washington, D.C., as a young professor, Benes saw something odd in the few preserved teenage brains in the collection: A strip of myelin seemed to grow bigger as teenagers aged.

Myelin, part of the white matter of the brain, is a cozy blanket of fat that wraps the long axon arms that extend from neuron cell bodies. It acts like insulation, keeping the brain's electrical signals on their intended paths along the axons and increasing their speed.

As a brain develops, myelin is manufactured from glial cells, the so-called glue of the brain. Outnumbering the critical neuron brain cells ten to one in the brain, glia also act as garbage collectors, carting away dead neurons and underused dendrites. Long considered the poor relations to neurons, new evidence suggests glial cells may play a bigger role, perhaps helping to increase signals. (Einstein's brain, researchers once found, had more glial cells than normal in areas devoted to logic and spatial reasoning.)

One type of glial cell, an oligodendrocyte, looks a bit like an octopus, and when the time is right—when the nerve cell's axon gets thick enough, possibly because of use—it sends out tendrils that wrap the axon like a jelly roll, creating the myelin sheath.

Once an axon has its myelin coating, it's considered white matter and its signal becomes more efficient and considerably faster, from creeping on a highway at rush hour to racing at NASCAR. An electrical charge travels a hundred times faster on a myelinated axon than on an unmyelinated one, reaching speeds of more than two hundred miles per hour. Not long ago, my brother-in-law began having problems with balance; he listed to the right when he walked. It turned out that he had multiple sclerosis, a disease in which the myelin casing on nerve cells disintegrates. In his still-mild case, there was a small demyelinated spot in his cerebellum, the lump of brain at the top of the neck that's involved in a range of brain activities, including balance.

There's no question that myelin is crucial for efficient brain function. And years ago, as Benes had stared at the brains in the bank, she thought she could tell, even with her naked eye, that the myelin was still growing in a key area in the teenage brain.

This, of course, was not supposed to be happening.

"We knew at the time that myelin was important in early brain development," Benes explained to me later in one of the large, chilly, dark-paneled rooms at McLean. "The time when you start to walk and your hands get more dexterous had been correlated with myelination in the motor cortex. But we thought that, overall, myelination in the central nervous system was over by the time you were five or six years old."

Though her sample was small, Benes published what she had seen. She then set out, systematically, to prove what she'd found. In Boston, she gathered brains from nearby hospitals and compared levels of myelin in the area of the teenage brain where she had detected growth, the superior medullary lamina. And she found what she was looking for: The brains of teenagers were still being enveloped in myelin. In fact, the myelin jumped a whopping 100 percent during the teenage years.

The area where Benes found the teenage myelin growth works as a relay station connecting two crucial brain areas, the cingulate gyrus and the hippocampus. The importance of the hippocampus, a cluster of cells in the middle of the brain, is well known. It's one of the primary areas of the brain for sorting out new memories.

In the movie *Memento*, the main character has to constantly write down what he wants to do and who he'd just met—he even has important information tattooed on his body—because, although his long-term memory is intact and he knows who he is, his hippocampus was damaged and he can no longer remember what happened even seconds before.

A patient who became a neuroscience legend, referred to simply as H.M., originally showed neuroscientists the world over what the hippocampus did. Because of severe seizures, H.M. had an operation in 1953 that removed a section of his brain, including both of his hippocampi. The operation reduced his seizures but left him unable to recall recent events. Without his hippocampi, H.M. could not transfer new memories into long-term storage. He retained certain types of memory; he could learn automatic motorlike behaviors such as drawing a circle or riding a bike. He could remember events before the operation. But, like the character in *Memento*, H.M. could not remember with whom he'd just talked, once explaining that "every day is alone by itself."

THE other area of the teenaged brain that Benes found undergoing myelination, the cingulate, involves emotions. The back part of the cingulate sends fibers to the brain stem and spinal cord, which control some of our basic gut reactions, such as a racing heart, a sweaty palm—perhaps even the urge to slam a door.

In other words, the section of the brain that Benes found still myelinating during adolescence is an integral part of a circuit that connects quick reactions to historical, contextual thought. Her findings raise obvious questions: If that area is still being built during the teenage years, could this help explain some of the bumps and blips of adolescence? If the connections between gut reactions and intelligent responses are not yet wrinkle-free, could that be one reason an otherwise polite fourteen-year-old boy flips out when asked to, say, take out the garbage?

"What we experience as emotion has two components: a gut feeling and an idea," Benes explained. "If you see a family member or close friend you feel good because there's a gut experience that is associated with 'liking.' At the same time, you also see the person and have an idea about that person as someone you like.

"During childhood and early adolescence emotional experiences are not very well integrated with cognitive processes. That means you may get an impulsive action that seems to bear little relation to what is otherwise happening."

Benes found that girls' brains were generally myelinating faster than boys. This, she suggested, could be part of the teenage gender puzzle. It may be one reason why young girls often seem to attain emotional maturity before boys.

"This myelination of pathways, this gradual tightening that is taking place, could be one reason why teenagers become more capable of mature forms of behavior, of better impulse control, of better focus and attention," she said

"But in a way," she added, "it's too bad to lose all that, don't you think? Teenagers are full of exuberance, that's what drives us. We adults tend to keep it under wraps; we wait until we get home. Sometimes I think it's too bad we can't keep some more of that."

Crossing the Great Divide

In a cramped neuroscience lab at UCLA, a young man monitors three computer screens at once. As he clicks his mouse, a screen on his right lights up with the outline of a brain. In the middle of the image is a wide horizontal strip throbbing in brilliant red. Slowly, the red begins to fade. The narrator, outlining the lab's research on the teenage brain, explains that the red color represents the growth rate of myelin in the corpus callosum, a large band of fibers that connects the left brain hemisphere to the right hemisphere.

The brain on the screen is a composite of living, human brains that were scanned as they aged from seven to sixteen. Initially, the band of fibers is myelinating furiously, then it settles down; the red on the screen goes from bright to muted. Through the magic of computer simulation, I watched the brains of adolescents as they smooth their connections, iron their wrinkles, grow more efficient, faster, precise.

The work focused on the ongoing myelination of part of the corpus callosum that connects still more crucial areas in the brain—the left and right sections of Wernicke's area, a region that helps us comprehend language and speak coherently.

The corpus callosum, with 200 million fibers, crosses the brain divide between the left and right hemispheres and helps integrate what are often two completely different views of the world.

There are times, for instance when someone has severe epilepsy, that surgeons cut the corpus callosum to control seizures. Such split-brain patients, as they are called, come out of the operation surprisingly well, but exhibit a few quirks that aptly illustrate what the corpus callosum does. (Very young children, whose brains are incredibly adaptable, can have an entire brain hemisphere removed, as a treatment for severe epilepsy or even tumors, and they often show few major deficits. If their left hemisphere is removed, for example, the right side will largely take over language functions, a marvelous feat that older brains cannot manage.)

With split-brain patients, the brain's right and left hemispheres are intact, but the bridge between the two, the corpus callosum, has been cut. Scientists, in particular neuroscientists Roger Sperry and Michael Gazzaniga, have measured the resulting quirks with some classic tests. Suppose, for instance, a split-brain patient sits at a table that's divided in half so that his left eye can see only items on the left side of a screen and his right eye only those on the right.

If a picture of a spoon is flashed on the left side of the screen, the patient will see it only with his left eye, whose connections go to the right side of the brain. And because the right side of the brain has few word-identifying areas, he's likely to tell you he sees nothing at all. He cannot put a word label on what he sees, so his brain decides it saw nothing.

But if you then ask that person to take his left hand, which is controlled by the right side of the brain, and pick out—from an assortment of objects under the table—what he'd seen, something interesting occurs: He will invariably pick out the spoon. The right hemisphere, which couldn't name the spoon, could nevertheless recognize its shape, a more tactile function, and pull it out from under the table.

Such split-brain patients have been known to try to take off clothes with one hand and put them on with the other, or pick up a book that looks interesting with one hand and then put it down after the other side of the brain, which can't read, decides holding a book is boring. In short, without a corpus callosum connecting the two sides, full brain communication is hit or miss at best.

And if, as the UCLA film suggested, the two sides of Wernicke's, a major language area in the brain, are not yet fully tied together across the vital corpus callosum isthmus during parts of adolescence, what does that mean?

Two young neuroscientists, Paul Thompson and Elizabeth Sowell, helped to explain. Thompson, dark-haired and boyish-looking, is from Leeds, England, a graduate, in Latin and Greek, of Oxford, before getting his Ph.D. in math. Sowell grew up in Southern California, and stumbled into brain-scanning out of an interest in psychology, getting her Ph.D. at the University of California at San Diego, a highly regarded center for neuroscience. Both had recently published papers in scientific journals, and the presentation I viewed at UCLA on teenage brain development had been based on their findings.

Thompson's study, the basis for the myelination film, had found that the fibers of the corpus callosum that connected Wernicke's area on the brain's left side with its counterpart on the right were myelinating like "wild fire" before dramatically calming down as kids moved from early adolescence through their first teen years.

What's so important about improved connections within Wernicke's? It once was thought that there were only two main language sections in the brain. One, Broca's area, located at the hairline roughly over the ears, was discovered by Paul Broca in 1861 when he did an autopsy on a man who, before his death, could say only "tan." Those with damage to Broca's area have difficulty finding and producing words. In my house, we've had ample evidence of what this region does. My husband, Richard, has a small knot of blood vessels, a cavernous angioma, near Broca's area that occasionally acts up, leaving him, for a few minutes, able to say only one or two ridiculous and random words, such as lamp or car. But even during these rare episodes, Richard is fully capable of comprehending language. If you tell him to sit down, he will. That's because his other main language area, Wernicke's, is still humming along nicely.

Wernicke's area, a bit farther back in the brain, was discovered by Karl Wernicke in 1876 when he studied a patient who could understand language but spoke only gibberish. Scientists still believe these two brain regions, Broca's and Wernicke's, are crucial language areas, but over the last several years they have identified nearly one hundred areas of the brain for different aspects of language, including special regions for the names of vegetables or tools. Even Broca's and Wernicke's are now thought to be subdivided. Some parts of Wernicke's, for instance, are devoted to comprehending emotional aspects of language, such as tone, and others help with syntax, such as the difference between Bob loves Sally, and Sally loves Bob.

Generally, the left side of Wernicke's, more dominant in most right-handed people, is thought to be more important for hearing words and figuring out what they mean. Its mirror region on the right side deals with the big picture, helping with such things as prose style, organization, or the ability, as Thompson put it, to "write like Emily Brontë." Although such skills obviously improve with practice, the ability to smoothly shuttle information between these two sides is considered essential for consistent higher-level functions of language. Thompson believes that improvement is linked to myelination of language areas in the teenage brain.

"A ten-year-old writing an essay on what he did on holiday might do it in a straight telegraphic style, 'I went here,' " says Thompson. "But in later teens, that same essay is likely to be more emotional, more organized."

And the changes involving myelination in the adolescent brain do not end with Wernicke's and speech. Thompson recently found that the area of the corpus callosum that sends fibers to the parietal cortex, the part of the brain linked to logic, doesn't even begin to myelinate, or "beef up," until about age seven.

"The parietal cortex is specifically activated when you do math or logical thinking or even crossword puzzles," says Thompson. "And there appears to be this bulking up of the white matter in that area in puberty. It makes sense. We don't teach algebra to little kids."

Tomas Paus, a young Czech neuroscientist at the Montreal Neurological Institute, in a recent study published in the journal *Science* that used 111 brain scans from Jay Giedd (who shares his teenage brain scans with scientists around the world), found still more critical brain areas that were still myelinating during adolescence. One, called the arcuate fasciculus, forms a link between the two important speech areas—Broca's and Wernicke's—and another area hooks together regions known to play a key role in fine-motor movements, such as the ability to touch type a letter or rapidly tie a shoe.

No one is sure yet whether this progressive myelination is due to a preprogrammed genetic plan or is pushed by use. (In other words, if you talk in more complex sentences does fat grow on your arcuate fasciculus, or does the myelinating fat grow first, and as a result those sentences start to sing.) But many neuroscientists themselves have been surprised to find such a level of change in the fundamental architecture of what was, not long ago, considered the largely "finished" adolescent brain.

"No one ever thought that what was changing (in teenagers) could be driven by, or be a reflection of, the basic changing structure of the brain. That is very, very new," Paus said.

Myelination in the developing teenage brain, however, has both good and bad consequences. After neurons are completely myelinated they are more efficient and faster. But there's a tradeoff. They also become more rigid. This may be one of the main reasons why younger children, as Thompson at UCLA says, seem to soak up foreign languages—and learn them without accents—considerably easier than older adolescents or adults. Once the brain's language areas are fully myelinated they are more specialized, considerably more sensitive to the sounds of the language that has been most frequently heard and less attuned to other foreign sounds, a fact that led neuroscientist Harry Chugani to wonder: "Who's the idiot who decided that youngsters should learn foreign languages in high school?"

Thompson and Sowell think some of the increases in gray matter volume that Giedd found at puberty might be increases in glial cells gearing up to myelinate. It may be that the teenage brain is also undergoing, as Thompson said, an "exuberance of glia." And with that exuberance, with the smoothing of connections of key brain areas, may very well come a clearer path out of the fog of adolescence.

"It's sort of like knowing five ways to get from Sherman Oaks to Westwood and being confused as to which one to take, and then, after you find the most efficient route, the pattern is reinforced and myelinated and set," Sowell said.

"Getting It"

There is little question that those in the non-neuroscientist world—parents, teachers, even kids themselves—recognize both this fog and the gradual clearing as well. In areas as diverse as writing an essay, remembering to call home, turning in homework, picking up a wet towel, or even learning to serve a volleyball or type a term paper on a computer, there is, if luck is in the wind, a growing mastery, a growing ease that develops in the teen years. Puzzle pieces scattered on the table come together, the picture makes sense.

Stanlee Brimberg, who for many years has taught seventh and eighth graders at the School for Children at Bank Street College in New York, says it's clear that individual experiences matter, family situations matter, a kid's temperament matters, good teaching matters.

And it also appears, he says, that something else quite fundamental is also taking place. Sometimes, he says, in the space of a year or so, kids just start to "get it" in a whole range of areas.

"If it were just one piece, if that kid suddenly got good at doing homework or something, then I would say it was just experience that did it," he said. "But sometimes all sorts of things they didn't get earlier start to click. They can pick up after themselves;

they can do their homework; they can answer questions without digressing, all of it, the spatial and the temporal, seems to come together and, I have to say, somehow it seems neurological to me, like some synapse closed or something."

Wendy Quirk, who has taught math in a suburban school district for more than thirty years, says that over and over she watches kids come out of their mathematical fog and "get it" when faced with the abstract concepts in algebra. Some kids, she says, seem to be born with an innate quantitative sense, but most others follow a similar timetable. "For most, it begins in the sixth grade, and then there can be this quantum leap," she says.

By the end of sixth grade about 50 percent of the kids, she estimates, have moved securely from the "concrete" to the beginnings of abstract, symbolic thought. To catch those who have not reached that point, she uses tangible examples to explain concepts. "I'll make one kid pretend he's x and another kid will be y and I put them in the front of the room and make them act out the equation," she says. "They need to see it."

By eighth grade, such antics are less needed. By that time, she says, nearly 80 percent of kids have some firm understanding of the abstract concepts in math.

So where does that leap come from?

"It comes from good teaching; it comes from practice—they have to drill—it comes from having good brains," Quirk says. "But I also think it comes from brain development. In the teachers' room we talk about kids who just aren't 'there' yet. But you know what, a little later they are. It's fabulous to watch."

In other more emotional and intellectual areas, the getting "there" also seems, to many parents, to come in stages. Connie, the mother of a bright fifteen-year-old girl who has been highly verbal and analytical for years, says she was nevertheless amazed to watch this ability expand exponentially as her daughter moved through early adolescence. Connie has never given a moment's thought to whether her daughter's super luminary medulla was myelinating, but she has watched her growing capacity to link emotional and intellectual worlds.

"I remember that before middle school, if she was talking about her day, it was more about events. It was more primitive. Speaking of other students, she would say, 'She was mean or he was bad,'" Connie said. "Later on, after elementary school, you start to hear things more like: 'She was mean because she had a fight with her mother' or 'Julie brags about her sister's SAT because she's so insecure and I guess she has something inside her that makes her do that.' They begin to see the invisible links between emotions and events. They see what's not there."

Connie believes her daughter learned a lot of that from listening to her "overly analytical parents." But even though her daughter may be prone genetically—and environmentally primed—to engage in such behavior, she says, "It wasn't there when she was eight years old and, when she was older, it was. It's progressive and that makes me think it is something in the brain."

What seems clear to neuroscientists like Thompson at UCLA and many others is that what many parents and teachers have noticed for years is, in fact, true. The teenage brain undergoes changes in nearly every area that scientists have looked at, including language and motor control and impulses. Such changes, Thompson says, must inevitably be connected to how teenagers relate to the world, and how the world relates to them. "The changes in the brain must have functional consequences," he says. "Teenagers process language differently, perhaps they assess risk differently. We know those are functions of the systems that are changing. Adolescent brains ramp up."

GOOD-BYE TO THE GRAY

At Thompson's UCLA lab, I watch a second digitized version of another human brain. This computerized composite portrait also represents the changing brains of a group of teenagers, but the group is older, from age sixteen onward. And, as the movie makes clear, it is a time when the adolescent brain is no longer spending so much time growing tangles of new neuron branches as it is working furiously to cut back whole forests of them, particularly in sections of the frontal lobes. As adolescence progresses, science is now finding, there is an enormous loss of gray matter—a wholesale slashing of cell branches and synapses—that goes on in the teenage brain.

While Jay Giedd, working in Washington, had found the gray matter growth spurt in the frontal lobes at the early edges of puberty, Thompson's UCLA colleague Elizabeth Sowell found a striking gray matter loss after age sixteen, particularly in the frontal lobes. According to the latest estimates of Sowell and Thompson, the average teenage brain cuts back 7 to 10 percent of its gray matter between the ages of twelve and twenty, with some smaller regions losing as much as 50 percent.

Thompson recently found that a small structure deep in the brain, the caudate, a motor-control section, prunes nearly 20 percent of its gray matter during early adolescence, first blossoming from around eight to eleven years of age, and then reaching mature levels, after a massive loss of tissue, around thirteen.

"The caudate is well understood; it governs the unconscious, mechanistic motor movements, the kind that become automatic once you learn them, like playing the piano, riding a bike, gymnastics," says Thompson. "I think people who teach those skills would probably agree, if you do certain things early on, if you excite the cells in this area, it may very well be relative to keeping those skills later on."

This thought, in fact, rang true to a number of physical education teachers I spoke to. Although none of them had heard specifically of caudates or brain pruning, the idea that young children should spend as much time as possible moving their muscles in a variety of ways, with certain skills getting cemented in by early adolescence, was hardly new to them. Ron Edward, who has taught physical education and coached young children and teenagers for twenty-four years in Arizona, says that it's obviously still possible for adults to learn new physical skills. But he considers it most beneficial and efficient for kids to "lock in" motor skills of a wide range of activities, not just one sport, "early on" so that they "develop a well-rounded, less rigid athletic body." José, who has played soccer since he was sixteen and is now twenty-three and on a competitive team, says he, too, has seen that kids who learn the sport from ages six to twelve are often considerably more facile, "more natural players."

Leaner, Calmer, and Quieter

At a certain point in adolescence, after the brain has reached its flowered peak, it begins, as Elizabeth Sowell's research confirms, to cut back, scale down, specialize. The teenage brain, all limbs and a bit loose, begins to pull in, quiet down. It sculpts a more specialized, more manageable, more manicured garden—from the jungles of Borneo to bonsai.

In fact, many of the connections that disappear during adolescence, scientists are now finding, are those that excite the brain and fire it up. Neurotransmitters, the chemical messengers emitted from a neuron, can have a number of effects on neighboring neurons, including jazzing them up or calming them down. The inciting of neighboring neurons is often accomplished by the brain chemical glutamate. And if glutamate-spraying synapses are disappearing in adolescence, as most neuroscientists now believe (by some estimates, the ratio of excitatory synapses to inhibitory synapses decreases from seven-to-one to four-to-one during adolescence) then, in some fundamental way, the teenage brain, too, is quieting down. Or, as Thompson puts it, "The disco music is being turned down a bit."

This is a concept that is well known to most parents. Ellen, a mother of twin boys in New York, says her sons got "very, very loud" when they hit adolescence. But she has been astonished, she says, by how both of them, now in college, have become so subdued. She can't judge what's happening inside their brains, of course, but viewed from the outside, the transformation has been startling.

"Now they will come and pat me on the shoulder and tell *me* to calm down," she says. "Who would have thought?"

Adolescents themselves, looking back, often recognize an evolution in their own behavior. Jessica, a senior in high school, admits she is astounded when she looks at freshmen who seem to be "running all over the place, yelling and screaming and fighting." Yona, another senior, remembers, as a freshman, how she would have wild thoughts of getting her whole body pierced, an idea that, looking back from age seventeen, seems to her now "silly" and "completely unnecessary."

IN carefully controlled work in their labs, neuroscientists are starting to pinpoint what the pruning process accomplishes in the developing teenage brain. Already, from watching where the pruning takes place, they know that it's connected to the fine-tuning of important brain functions, including inhibition control and working memory, or the ability to hold information in your head when there is competing information. Both monkeys and humans improve in those areas—the accuracy and the precision with which they can remember where a light flashed on a screen, for instance—along a trajectory that parallels the pruning of synapses in the prefrontal cortex. So while an exuberance of synapses might be important in acquiring these kinds of skills, it may only be with the pruning process that takes place in later adolescence that those skills are refined, perfected. According to David Lewis at the University of Pittsburgh, most people do not reach adult capacities in this type of task until the late teens.

In fact, scientists like David Lewis, who studies the intricacies of brain activity in adolescent monkeys, say that even the wide range of recently uncovered changes will not, in the end, prove to be the full story of the dynamic teenage brain.

In his most recent work, Lewis, along with several others, is finding that the teenage brain undergoes additional subtle alterations in markers of the effectiveness of inhibitory connections at specific synapses in the cerebral cortex, as well as what now appear to be shifts in certain vital neurotransmitters, the chemicals that let brain cells communicate with each other.

Lewis says there are major changes in the levels of dopamine, which rises to peak levels in the prefrontal cortex during adolescence and then declines before stable adult levels are achieved. Dopamine, one of the key neurotransmitters, often acts to modulate other neurotransmitters. And if dopamine levels are too low or too high—and dopamine has also been shown recently to increase with stress—those signals can be off.

In fact, the changes in dopamine may continue after the major pruning of synapses in the prefrontal cortex is complete. This means that long after the main construction project of the teenage brain slows down, changes in dopamine may continue to shake things up.

"No one has any difficulty distinguishing the behavior of a four-year-old from a ten-year-old," Lewis says. "But the brain is clearly changing later, too, and there are associated changes in behavior. I think you can say that the circuitry is being refined in a way that enhances both precision and capacity."

Elizabeth Sowell believes there must be some evolutionary reason why the brain remains so flexible during the long stretch of adolescence and then, only at the tail end, starts to consolidate. If it were totally organized and set in its ways too early, human beings, she points out, would never be open to learn the things they need to survive—wherever they may find themselves.

Developing Emotional Brakes

Patrick Russell was thirteen when he first went into the brain-scanning machine at McLean Hospital, in a scientific trial that looked at teenage brain connections from a different perspective. The scanner was set to take pictures quickly and measure not structure but levels of oxygen used as the living brain actually worked at various tasks.

The brain is an energy hog. Although it's only 2.5 percent of the body's weight, it uses 20 percent of the body's energy. Most brain-scanners measure the structure of the brain. But scanners can also be rigged to gauge the oxygen level in the blood that a brain absorbs when it is in action. When set this way, the scanner is referred to as a functional magnetic resonance imaging machine, or fMRI.

Getting clean results from fMRI work is even trickier than with regular MRIs, which are complicated enough. Still, many neuroscientists say they believe that, in the end, watching the brain as it functions in real time will prove the only way to truly figure out how it works on a minute level.

In fact, the experiment Patrick participated in had stunning and unexpected results. When Patrick and other young teenagers were shown, while in the fMRI, the face of a man in fear, the part that lit up most in their brains as it worked to make sense of that face was not the frontal cortex, the part the adult brain uses to sort out the complex nuances of emotions. Instead, the young teenagers' brains lit up in a section called the amygdala, an almond-shaped knot in the middle of the brain that is one of the key areas for instinctual reactions such as fight or flight, anger, or "I hate you, Mom."

Earlier, the McLean researcher who performed the test, Deborah Yurgelun-Todd, had done the same experiment with adults. And when the adults were shown the fearful face, their brains, as expected, dutifully employed their frontal cortex, the part that sorts out the various whys and hows, the part of the brain that says "Wait a minute, should I be afraid? What is that anyhow? Have I seen that before?"

So why was Patrick's teenage brain acting differently? Some scientists, including Yurgelun-Todd, believe he was processing emotions in a different part of his brain because connections in his frontal cortex were not yet fully wired up. And if teenagers respond to such an important social cue as a fearful expression with their brains' primal emotional center rather than their rational center, could that also help explain why teenagers can often seem to overreact, emotionally erupt for no apparent reason?

Yurgelun-Todd thinks the answer is yes. Adults respond with the part of their brains, the frontal cortex, that helps them to halt impulses, apply emotional brakes, and reason logically, while teenagers often don't.

Her work, she cautions, is far from definitive. Still, in another study, Yurgelun-Todd found other intriguing differences: Younger teens, in particular, not only tended to use the more crocodilelike amygdala part of their brain when they processed emotions, but they often got the emotion wrong to begin with, labeling fear as anger, for instance. They misread facial expressions; they got mixed up.

Another recent study also detected evidence of what might be called the teenage mental muddle. Neuroscientist Robert McGivern and his colleagues at San Diego State University found that as children hit puberty, around age eleven or twelve, the speed at which they can identify emotions drops by as much as 20 percent. Their reaction times remain slow for several years, returning to normal levels only at age eighteen, a finding, McGivern said, that could reflect the "relative inefficiency in frontal circuitry" of the adolescent brain as it undergoes a remodeling—its growth spurt and pruning of synapses.

Such discoveries raise the very real possibility that teenagers simply see the world differently from adults. Lacking sufficient experience that would help them correctly sort out social cues and lacking a fully developed and functioning prefrontal cortex that can provide context (perhaps she frowned at me because she is having a bad-hair day or because her boss yelled at her and not because she hates me), they may not always get it exactly right.

"We have to think about the idea that they might not be hearing the words in the ways we intend them," Yurgelun-Todd said.

THIS, too, of course, is not an unknown concept to parents. A friend, Susan, the mother of two teenagers, says that at a certain point at the beginning of adolescence, her son, an otherwise quite intelligent boy, "seemed unable to listen."

"If I say take six dollars and go to the supermarket and buy this and buy that, and then go to the hardware store on the way back, I know he will only do one of those things. So now I just tell him to go to the store and get this one thing," she says. "I don't know where it comes from. Is he just thinking of other things?"

Yurgelun-Todd says her findings have made her change how she relates to her own two young adolescents. As teenagers begin to look like us, there is a tendency to expect them to act like us. But that doesn't always work.

"I used to ask my daughter to put a dish in the dishwasher and brush her hair and pick up her clothes, and then I'd get angry that she only did one of those things," she says. "Now I don't expect her to hold so much information in her head. She does one thing and I expect that now. You look at some fourteen-year-old, maybe he's a junior-varsity football player and he looks big and mature. But his prefrontal cortex is not fully mature. People think that in high school, all we have to do is stuff the right information into them and they'll make all the right decisions. But their brains are functioning differently."

For his part, Patrick says he knows his brain has changed as he has progressed through adolescence. During the three years his brain has been scanned at McLean—a total of five times—he says he has grown calmer, more directed, more focused, although he's not exactly sure why or how.

Now sixteen and a junior in high school, he plays the euphonium (a small tuba) in the marching and concert bands and has developed a firm love of "all kinds of music." He notices that while he still gets angry (particularly when other kids in the band don't practice their music) he no longer "flips out" or feels a need to "punch things" like he used to. "I've learned to control my emotions better," he said.

Over the past three years, he has clearly grown outwardly. When he started his brain scans at age thirteen, he jokes he was "about four feet tall." Now he is a gangly five-foot-seven.

His father, Keith, who has saved copies of his son's brain scans, agrees that Patrick's body is not the only thing that has grown. "It's amazing, but if you look at the brain scans through the years, even I can tell the difference," Keith said. "Patrick's brain has gotten more stuff in it. It's more . . . it's more . . ."

Patrick, his brain connections smoothing out, speeding up, quickly filled in his father's blank. "Complex?' he offered.

"Yes, that's it," said Keith. "Over the past three years, you can see, even I can see it in the brain scan, Patrick's brain has gotten more complex."

BIOLOGY AND BEHAVIOR

Many neuroscientists acquired their belief in brain development and behavior quite naturally. Tomas Paus in Montreal, for instance, says that in his native Czechoslovakia scientists always gave more credence to biological psychology than to long-term Freudian psychoanalysis, so he grew up with the idea that what was happening to the brain itself was intimately tied to what was happening to the person on the outside.

But for others, the idea of linking brain biology and behavior in teenagers or anyone else has evolved only recently, even among those who are some of its chief advocates today.

Early in his career, after graduating from medical school in North Dakota, Jay Giedd had gone to work at the Menninger Clinic in Kansas. At Menninger, Giedd worked with a wealthy young man who had suffered for years from obsessive-compulsive disorder (OCD), in which people feel compelled to perform ritualistic behaviors such as tapping their fingers in patterns, or repeatedly washing their hands, to reduce fears. The man had spent years trying to get well, probing his past, Freud, toilet training, all that, Giedd said, but he got no better.

After working with that patient for a time, Giedd was sent to work at a Menninger satellite clinic in Phoenix and wound up helping in emergency rooms, where a young cocaine addict was admitted with extensive brain damage from a .22-caliber pistol he'd used in a suicide attempt. The surgeons saved him and he functioned well, but afterward the man developed OCD, the same as Giedd's patient at Menninger.

"Here was a man who was fine before and then, after physical damage to his brain in a specific spot, he had OCD," said Giedd.

Then Giedd heard that his former patient at Menninger had tried one of the new antidepressants, like Prozac, that adjust levels of serotonin in the brain and, within weeks, he was better.

"I had a kind of epiphany," Giedd told me. "Maybe all this humanness is really about what goes on in the brain," Giedd told me. "I still believe the psychodynamic model has a lot to offer. I use its principles in my role as a child psychiatrist. But I began to think maybe it was time to go look for the biological basis of behavior. Believe me, I wasn't looking for this idea. I was on the other side, but it just seemed so clear."

Not long after that, he took the job at the National Institutes of Health, joining a brain-scanning team being run by Markus Krusei, a child psychiatrist now at the Medical University of South Carolina. The goal of the NIH project was to map the developing brain of kids with psychiatric disorders, and Giedd began his research by trying to learn all he could about the normal development of child and adolescent brains.

He found next to nothing. There had never been a systematic long-term study of the brains of normal children, in part because of ethics. A good argument can be made for doing brain-scanning research on those with disorders—science is helping those who need it most. But the issue of whether the government could, in good conscience, repeatedly expose children with no problems at all to powerful MRI scanners was considerably thornier.

Brain scanners, or MRIs, unlike X rays, do not use ionizing radiation that can damage DNA. Instead, they work with radio waves that interact with hydrogen atoms in water molecules in the brain. When a person is put in a scanner, its powerful magnetic field first lines up the hydrogen atoms in the brain. Then, a jolt of radio waves makes the atoms bobble, and then fall back into place. The MRI measures the energy coming from the atoms as they return to their normal position. And when the calculations are fed into computers, the machine produces an outline of brain structures or, as Giedd puts it, "an exquisitely accurate picture of the living growing human brain."

The process, according to Giedd, has "no known effects" on the functioning of cells. Studies of MRI technicians who are exposed to the magnetic field far more than patients do not show increased health risks, he said. But it was studies of female scanning technicians who worked throughout their pregnancies and did not show any increased birth problems, he said, that were finally instrumental in convincing the NIH Ethics Review Board that the procedure could be used for normal children. And so, in 1991, the NIH began the first long-term brain-scanning study of normal children. In effect, the use of MRIs, Giedd said, "launched a new era of adolescent neuroscience."

Steve Hyman, now provost of Harvard, who headed the National Institutes of Mental Health while the push for normal brain studies gained momentum, says such studies were inevitable. With recent improvements in noninvasive scanning tools, as well as new computational approaches that allow comparisons of widely varied brains, he said, the time was right.

"Sometimes in science we like to heroically say that new ideas drive progress," he said. "But sometimes new technology is a very important contributor."

Today, in fact, the NIH is involved in the largest study of its kind, a $16 million, seven-year project to regularly scan the brains of five hundred children from the ages of two weeks to twenty-one years, as well as test cognitive functions and, through cheek swabs of DNA, examine their genes.

FROM his office at McLean Hospital, Nick Lange, a biostatistician, is in charge of eventually corralling all the brain scans into graphs that will, for the first time, map the normal growth of the human brain from childhood through late adolescence.

"You know those charts for arm and leg growth that every pediatrician has?" he said when I went to see him as the project was just getting off the ground. "We are going to have charts like that, but for the development of the human brain."

In the end, Lange said, "We will have an atlas of the developing human brain, one that will be able to tell researchers and doctors just how the brain of an average, normal fifteen-year-old is supposed to look."

In many brain studies, there's been a nagging problem. Unless you look at the developing brains of the same kids year after year, as Giedd did on a smaller scale, results can be confusing. Brains may look alike on the surface but be quite different. The brains of two girls, both thirteen, with the same IQ can vary as much as 50 percent in the size of certain regions.

What researchers desperately need, Lange said, is to find out what the normal range is so they can recognize deviations. They need to be able to look at the brain of a "fifteen-year-old who is having mood problems and see if the problems are in structure, maybe big problems linked to the prefrontal cortex, or it's just plain moodiness and we have to look somewhere else."

Understanding the behavior of teenagers, of course, is never going to be a simple enterprise.

"To fully understand a normal teenager," Steve Hyman says, we're going to have to know much more about "mood regulation," the delicate dance of outside influences and inside tugs and pulls, the varying influences of nature and nurture that change "periods of irritability to periods of optimism." As he points out, you can put "the same biological organism in Rwanda or Gaza or the Midwest and it will come out with a whole different set of beliefs" that will then drive a different set of behaviors.

In illnesses such as schizophrenia, thought to have a large genetic component, the identical twin of someone with the disease has a 50 percent chance of being similarly afflicted. But that means they also have a 50 percent chance of being perfectly fine. Is the key environment? As Hyman says: "Where does that other 50 percent come from? We just don't know the boundaries yet."

Still, he says, the new brain findings are crucial first steps toward figuring out why a teenager acts as he or she does, what makes them so similar to each other and yet different from both children and adults.

"Why is it that an adolescent is more vulnerable to tobacco and alcohol and addiction? Why is it that, if they start earlier, they get more addicted? Why do we see so much depression appearing and why more in girls? Why does schizophrenia occur? Why do some kids get derailed by learning disabilities that become prisons? Why do some boys form antisocial groups and go on to criminal careers and others begin careers in scholarship?

"And why is it that the brain seems set up to develop empathy?" asks Hyman, who has a eleven-year-old daughter he calls a "practicing adolescent."

"Why does the brain change in its sense of relatedness to other people? All that comes together in adolescence. Why is it that an eight-year-old sees his obligations to the people and the world around him so much differently from a fourteen-year-old? Not that a fourteen-year-old is fully mature, but he is certainly different."

While much of the science remains as raw as an average teenager at the breakfast table, Hyman says that the new brain work is helping to piece together a fuller picture and will help even more down the road.

"It's clear there's a burst of brain maturation, combined with learning and experience, that happens in adolescence," he said. Eventually, scientists want to be able to learn enough to intervene, to help when there are problems.

"But first," he says, "we have to get a baseline."

Chapter 6

The Adolescent Animal

From Chimps to Chekhov

When Fletcher reached adolescence, this is what he did: Deep in the night, he sneaked away from his home in Maryland, ran six miles to the Potomac River, and swam to a small island. He ate what food he could find, begging sandwiches from river fishermen. A search party looked frantically and finally found him huddled on the island—hungry, dirty, chagrined.

Fletcher is a rhesus monkey and his story is a favorite of Steve Suomi, head of the National Institutes of Health animal center in Maryland, whom I went to see one hot afternoon to talk about adolescence of a different sort: puberty à la primate.

Animals—from rats to rhesus monkeys—have their own form of adolescence, a distinct period between childhood and adulthood that may represent the roots of our own protracted adolescence. And to parents of teenagers, the animals' behavior during that period might prove surprisingly—reassuringly, disturbingly—familiar.

As I spoke with Suomi at the animal center, we stood outside

neural endocrine system, with similar hormonal changes, they, too, go through a "pronounced growth spurt, when they grow like crazy," just like human teenagers. They show clear cognitive growth and, as they progress through adolescence, get better at abstract thinking, just like their human counterparts. In prepuberty, a rhesus monkey has a hard time with what's called the "oddity test," which measures abilities to detect the difference between a square and two circles.

"Before they can't get it, but at puberty they can . . . as long as they get a Froot Loop," said Suomi. "They're very much like teenagers. They have a similar pattern of brain growth—they get more frontal connectivity with other parts of the brain like the amygdala [an emotional center of the brain]. They can control things better."

Females reach puberty at the end of their third year and males at four. Before puberty, during monkey childhood, they spend lazy stretches hanging out with friends. At around six months, the average rhesus monkey begins to spend only 20 percent of its time with its mother and prefers to romp around with monkeys its own age. That's the time, Suomi said, when young monkeys learn about hierarchies, power struggles, and strategies for success, such as forming alliances and persuading relatives to help them in fights. "Ah," I said, "middle school." Suomi laughed, agreeing. "It's very much like a small town."

At puberty, a split occurs. Female monkeys abruptly stop playing with friends and retreat to mothers and aunts, learning to groom and care for babies. Males leave the troop, voluntarily or, if they're obnoxious enough, older females kick them out. Teenage male monkeys form juvenile gangs, foraging for food and fighting, before joining other troops, a pattern that's repeated in many mammals, probably to avoid inbreeding. This was, in fact, the driving force that sent Fletcher flying off to the Potomac.

It's at this point in monkey adolescent development, Suomi said, that inborn personality, as well as early experiences, can play a big role. More aggressive monkeys can get kicked out of a troop too early and die in the gangs or get killed while attempting to knock off a dominant male in another troop too quickly. It's sometimes the shy monkeys that do better. They're less annoying, older females tolerate them longer, and they're more mature before trekking into the wild. "You see," said Suomi with a grin, "sometimes being a nice guy does pay off in the end."

Still, there's a 40 to 50 percent mortality rate for male teenage rhesus monkeys once they leave. "It's a pretty rough world for male adolescents," Suomi said.

Leaving the outdoor compound, Suomi and I went inside one of the small buildings nearby and, after putting on hospital gowns and masks, entered the monkey nursery. Inside, were rows of cages and tiny gray monkeys sucking on bottles sticking out of terry-cloth-covered poles—their surrogate mothers.

I thought they looked a bit sad, hugging those poles, but Suomi said that, in fact, monkeys raised by terry-cloth mothers do better than monkeys who grow up only around friends. A protégé of Harry Harlow, whose work showed the devastating impact of depriving monkeys of contact with their mothers, Suomi and his team have continued that work, extending it to find out how those bad effects, stemming from poor parenting or unfortunate genes, can be reversed, if at all. Suomi said that about 20 percent of rhesus monkeys he has studied are born shy. Their shyness is measured by their behavior—they are easily frightened—but also by careful measurements of the blood levels of the stress hormone cortisol. (The percentage mirrors what Jerome Kagan at Harvard has found studying the number of naturally shy young children.)

About 10 percent of monkeys are more impulsive and aggressive than others. "They do stupid things and get in fights a lot," said Suomi, who added that they also tend to have lower levels of serotonin, the neurotransmitter that can have a calming effect and is the basis for antidepressants like Prozac. The impulsive monkeys have few friends because their play escalates into aggression. They fail to learn the social niceties, irritate the old ladies, and get the boot early. Another 20 percent of the monkeys are prone to depression. "They roll up in a ball and cry when their mothers go off to mate," Suomi said. Their heart rates and levels of certain neurotransmitters are higher than normal monkeys, and they get better with antidepressants.

Watching these patterns, Suomi said, it's clear that genes and early parenting can determine a great deal later on, in adolescence. But, he said, that outcome can also be altered along the way. His work has consistently shown that good parenting—that usually means a consistent parental figure—can alleviate some unfortunate inborn traits.

Monkeys raised by nurturing mothers or even with terry-cloth mothers fare better than those raised only around friends. Peer-raised monkeys, Suomi said, form weaker attachments and, although they look fine physically, are more uptight and reactive to stress, particularly by the time they're in adolescence, when stresses increase. And, he said, given the opportunity, they also "tend to drink heavily."

Suomi, like all those who study primates, is cautious about ascribing these kinds of findings directly to teenaged humans, with complicated brains and intricate high school politics. "Monkeys are not just little humans with furry tails," he said. Still, studying monkeys can offer insights about what it means to go through adolescence as a human. "We can see patterns of development that have meaning," said Suomi, who said he himself had been a naturally shy teenager who managed to overcome that genetic predisposition through "support groups" he developed through interests in sports, music, and eventually monkeys.

His point is that early experiences, good and bad, can alter brains and have an enormous impact on behavior later on, particularly in adolescence. Certain types of early experiences, he believes, can have lasting effects not only on a brain's physical structure but also the finely tuned functions of neurotransmitters. Monkeys that have an unfortunate genetic makeup or poor nurturing or both often seem okay at first, but, if they aren't helped in some way as they head toward adolescence, the problems can become more pronounced.

"At that point," he said, "they're trying to do things in a

grown-up world and they get in trouble because they do things poorly."

WHEN MATING TIME ARRIVES

Anne Pusey, a behavior ecologist at the University of Minnesota, sees patterns in chimpanzees similar to rhesus monkeys, although with a twist. With chimps, for reasons that are still unknown, it's the female teenagers who leave the group. They're forced to seek mates in neighboring chimp colonies, a decidedly risky move since established females don't necessarily want the pretty young things around.

"I guess you do see this with some humans, don't you," said Pusey. "Think of the young girls in India or China who are forced to marry into families and then their mothers-in-law are in charge."

Pusey studied chimpanzees in the Gombe National Park in Tanzania for more than ten years. And chimps, too, she says, have a distinct period of adolescence, although it's difficult to measure in the wild. Females show the first signs of puberty, sexual swellings, around age seven, but they don't have big enough swellings to mate with adult males and generally don't reproduce until they're ten. Males start getting bigger testes around eight, but generally don't become fathers until around thirteen.

While chimps' growth spurts may be less pronounced than in humans, Pusey sees clear behavior changes that mark a chimp teenage period. "However you define adolescence, there are long periods when sex hormones are circulating and changing their behavior long before they are considered adults," she said. When puberty hits young teenage males, for instance, they become fascinated with older males and want to spend time with them, though initially they're terrified.

"I remember seeing one young teenage male chimp that was going toward the grown-up male group and he kept looking back at his mother to come with him. They like to be with their mothers," Pusey said. "And some of the mothers plod along behind."

Before they take the final leap to join adult males, however, teenage male chimps try out their growing strength in a disturbing way. Around puberty, male chimps, perhaps copying the aggressiveness they see in older males, begin to batter young females, dominating them to such a point that, later on, they can have sex with them whenever they want.

"It's disturbing and horrible to watch," said Pusey. "Sometimes you see them all sitting around grooming and everything looks fine. But there's always this underlying threat."

Although she sees parallels between chimp and human adolescence (aside, one hopes, from the battering), Pusey said there are still clear differences. One of the most startling differences is the lowering of the age of puberty in modern human adolescence. In part because of improved nutrition, puberty now starts fully two years earlier in human girls, and perhaps in boys, than it did a hundred years ago. At the turn of the century, young girls began to menstruate at age fifteen on average, and those averages are now down to thirteen. Pusey wonders whether, with the lowering of the age of puberty, we are dealing now with some kind of "disconnect" between the human brain and body.

"You do have to think: Is the brain keeping up?" she said.

The Birthplace of Intelligence?

Why do humans have such a prolonged childhood and adolescence in the first place? What exactly is the point of being protected and dependant on adults for a full eighteen years and sometimes considerably longer? A chimp, after all, can usually make it just fine on its own after about eleven or twelve years with mom.

Many evolutionary anthropologists think this outsized human phase developed because we have evolved such complex societies, with styles of hunting and foraging so complicated that it takes nearly a third of our lives to master them. Others more recently have disputed that idea, saying that in some still-primitive societies, children can perform sophisticated hunting techniques as easily as adults as long as the work does not rest on strength and size. Perhaps, these scientists say, our drawn-out adolescence has evolved only to balance our comparatively long lives. As we got intelligent enough to dig for the most nutritious tubers and began to live longer, it made sense to postpone reproduction until physical development and social position were more secure, and reproduction could be more often successful.

Under that scenario, it also might make sense for teenagers to look as young as possible for as long as possible so that, while the extended growth takes place, they don't look threatening to an average grown-up, who will then be more inclined to patiently pass on the ways of the world, instead of batting them away as menacing competitors.

Barry Bogin, an anthropologist at the University of Michigan at Dearborn, has come up with his own version of how this non-threatening, modern adolescence might have developed in humans and why. And he believes it is all tied to the human growth spurt, whose timing has helped make us the rabid reproducers and the successful species that we are. In Bogin's view, it is this lengthy and specifically timed adolescence that made us human.

As Bogin sees it, the adolescent growth spurt works differently for males and females for a good reason. Girls have their growth spurt earlier than boys. Around eight years of age, most girls get some stirrings of hormones and their bodies start to change: a wisp of pubic hair, tiny breast buds. About four years later, at an average age of twelve or thirteen, girls hit official puberty, called menarche, when they start menstruating. During that time, since they have begun to look like grown-up females, Bogin says, the real grown-up females take notice and take them under their wings, teaching them what they need to know, particularly how to keep a baby alive and otherwise be a successful reproducer.

But even after girls are menstruating, most don't become fully fertile until later. Their pelvic bones don't grow to a good size for successfully carrying a baby to term until they're about seventeen. And, perhaps more important, ovulation, as Bogin puts it, remains "hit or miss." Girls are not considered fertile until about 80 percent of menstrual cycles include ovulation. And that doesn't happen until they're about nineteen, something that has not been lost on cultures along the way.

"Around the world and through the ages—from colonial times to [present day] Guatemala—girls have their first babies on average at age nineteen," said Bogin. "That's just the way it is."

Bogin also believes that the gap evolved because our societies became more complex. As humans evolved, females needed more time to teach more complicated skills to their offspring. The lag time—between the time when nonfertile young girls looked sufficiently female so that older women included them in their social circle and the time they began to successfully reproduce—gave girls time to learn how to be better mothers. And it gave those early *Homo sapiens* who developed that learning period an evolutionary advantage.

In fact, humans are an enormous reproductive success story. While chimps manage to raise only about 36 percent of their infants, humans raise a whopping 60 percent. "Which is why the world is lousy with humans," Bogin said.

Boys, on the other hand, follow a pattern opposite that of girls. They become fertile long before their growth spurt. On average, boys have their first nocturnal emissions and produce sperm at age fourteen or fifteen, but they don't grow all that male muscle until sixteen or seventeen. That means that boys, primed by hormones, become interested in grown-man activities, interested in hanging around the hunters perhaps, while, Bogin said, they still look like "wimps" and aren't so threatening.

"They could do stupid things, but the older guys would just laugh at them instead of killing them," Bogin suggested.

Bogin concedes that this is only one theory, but to him it's the only one that makes sense. "Adolescence is fairly recent" in human history, Bogin said. "And it developed because it was a survival mechanism for the species."

But what does Bogin think of adolescence today? Clearly, we are far removed from our evolutionary roots. What's pushing us toward the long SAT and MTV–saturated state of adolescence we have now? Culture, it seems, has taken the stage.

"Everything that happens now with humans is biocultural," Bogin said. "A lot of things go into it."

INDEED, for most of the modern era, the only way to determine what makes teenagers do what they do has been to watch them from the outside as they flop about in that biocultural soup. Most developmental psychologists trace our modern view of a necessarily angst-ridden and endless adolescence to G. Stanley Hall, a professor of psychology and friend of Sigmund Freud's, who, in 1904, published a book proposing that teenagers were essentially primitive savages, overcome by emotion on a necessary but troubled passage to civilized adult behavior.

Other more recent attempts at understanding the teenager have left mixed messages at best. Patricia Hersch's *A Tribe Apart*, a richly chronicled tale of the lives of eight teens in Reston, Virginia, described them as a lonely, sullen bunch, "remote, mysterious, vaguely threatening." David Brooks, writing in the *Atlantic Monthly* about older adolescents entering Princeton—kids he considered the budding elite—found them like well-behaved robots. Having spent their lives being escorted to "skill-enhancing" activities, they are "bright, morally earnest and incredibly industrious . . . the organization kid."

Thomas Hine, in his wonderful and pointed *The Rise and Fall of the American Teenager*, traces the very word teenager to Madison Avenue, which in the 1940s needed a new marketing demographic. As Hine described it, the definition of a teenager has veered all over the map, an artificial social invention bent and twisted to fit the cultural needs of the moment. Young upper-class boys in Sparta had tutors who taught them to steal and intimidate the slaves. Indian tribes sent pubescent males to the mountains to await an adolescent vision, perhaps a dream of a bow and arrow;

teenage Indian girls, at the first sign of menstruation, were banished to a hut to be purified of evil spirits. Not that long ago, poor teenage girls worked in textile factories for fourteen hours a day and richer ones were loaned to friends as servants. More recently, fourteen-year-old boys were fighting in Afghanistan. Hine sees our current Westernized crop as little more than pampered mall rats serving "a sentence of presumed immaturity, regardless of their achievements or abilities."

"For many individuals such a long period of education, exploration and deferred responsibility has been a tremendous gift. For other individuals, it has not been a blessing," he wrote.

THE biggest attempt to understand teenagers from the outside looking in is the National Longitudinal Study of Adolescent Health that began in 1995 and regularly surveys ninety thousand teenagers in church groups and rock clubs. Robert Blum, a pediatrician and professor at the University of Minnesota who has analyzed much of the data, said it, too, has a mixed message.

Most teenagers—about 80 percent—traverse the teenage years with aplomb. But these teenagers' success, Blum says, isn't based on immutable attributes such as wealth or race, but on more mundane things like having at least one adult who cares about them and being connected to their school. Ann Masten, who has studied resiliency in teenagers as diverse as homeless kids in Minneapolis to Cambodian refugees, calls these simpler things "ordinary magic."

Most recently, however, there have been hints that teenage success or failure can be affected by biology. Some of the most consistent findings about teenagers over the past few years have been that those adolescents who mature off schedule run into the most difficulties. Boys who mature late, for instance, stuck with squeaky voices and nonathletic physiques, tend to have lower self-esteem. But both boys and girls who mature much earlier than their peers are more likely to use drugs and alcohol and have early sex.

Some troubling teenage statistics in the United States have improved in recent years. Teenage birth rates, while still towering above that of other developed nations, were at a record low in 1999, for instance. Tenth graders were less likely to smoke in 2000 than in 1996. Blum's survey found that 75 percent of teenagers were religious and considered their parents heroes. *Heroes*!

Still, Martha Erickson of the University of Minnesota, who has analyzed data from another longitudinal study, says adolescents today, dealing with powerful drugs, AIDS, and cafeteria and street violence, face "stakes that have not been so high since pioneer days." They remain, she says, a Jekyll-and-Hyde population, with some cracking the SAT barrier and others attending funerals of friends killed by gunfire. According to Blum's survey, one in four teenagers, or 5.3 million youths, has been involved in an incident with a gun or a knife in the past year; 20 percent of seventh graders and 60 percent of high school juniors had had sexual intercourse; 10 percent drank weekly. "The group seems to be split; there's this polarization," Erickson said.

AND into this anthropological-psychological-sociological and far from clear soup have plopped the neuroscientists. Can their new findings, their new equations, their new vantage point help figure out which teenagers will fly and which will flop? Can they explain why some gobble Ecstasy, while others feast on calculus? Bob Blum says the new perspective will help.

"What they [neuroscientists] are finding about the neural development of the adolescent brain is absolutely fascinating, even awesome" he told me. "Ten years ago there was nothing. Now I think it will be the frontier of the field for the next ten years. It will change the whole debate about adolescents. It will have huge implications for policy, for laws. It will change the whole way we think about kids. Forever."

Chapter 7

Risky Business

Why They Do the Things They Do

One girl down the road, at age fourteen, the one we used to watch in ballet class—the tallest, the prettiest, the most graceful—was caught sneaking out her window at midnight to have sex with an older boy.

The boy around the corner, age thirteen, the tousle-haired one who used to dazzle us with his high-wire skateboard antics—up on the rails, over the walls—was caught sneaking in his window at midnight after hanging out with older boys, drinking, and smoking marijuana.

Teenagers doing dumb things. It's an old story. But why do they do it? And why do some do it more than others?

Teenagers themselves, particularly the younger ones, are stunningly inarticulate when trying to explain why they do risky things. One boy, thirteen, spent an afternoon at one of those skateboard parks where he careened down one side of a concrete skateboard track, flew high into the air, turned in midflight and then careened down the other side at full speed. He had no helmet, an air of slight defiance, and said he did it only because it "was a lot of fun." Another boy, fourteen, who spends every minute he can on his mud-covered mountain bike sailing over cliffs and wide ravines, told me it was just something he "felt like doing."

Older teenagers often have some rationale for their risky behavior, however unconnected to reality. Jessica, seventeen, talked about the time, on a recent senior trip to the Catskills, when she was supposed to be a leader. Instead, she spent the afternoon smoking marijuana in the room next door to the adult counselors.

"Sometimes I think I take risks when I'm feeling vulnerable; maybe I want to forget that feeling and so I do something that will take my mind off it," she says. "Sometimes I want to test the limits. I think sometimes that's a lot of it. You think about the consequences, maybe a little, but then you defy them or maybe you convince yourself, for whatever reason, that those consequences don't apply to you because you have this need to do what you want to do."

An eighteen-year-old boy, a stellar student on his way to the University of Chicago, says he used similar thinking when, for a period of time, he regularly shoplifted small items from local stores. "You think, I'm good, nothing will happen to me," he said.

Another seventeen-year-old girl told me she had a need to live a "little on the edge." Like Fletcher, the rhesus monkey that was compelled to leave his troop and explore other territory, the girl said that she had to "get out there; find out about the world and what her own capabilities were." Sometimes that has meant walking, on purpose, and alone, at 2:30 A.M. in a crime-ridden part of town.

"It's exciting and thrilling and I find I like that; it's an adrenaline rush," she said. "I don't know what my limits are and I guess I want to find out; find out what I can do."

Her friend Vanessa, a poet, said that she had a need to take what she called "emotional risks. In high school it can be a big risk to walk on the wrong side of the cafeteria wearing the wrong thing," she said. "But sometimes I actually like doing that. I want to rebel against the way things are, like the world's concept of beauty, which I find too narrow. I want to do that just to see what happens."

A few things are known about teenage risk-taking.

First, while most teenagers do a few stupid things, only a few get into any real trouble. Most teenagers come out just fine.

In fact, a fair amount of risky behavior is not only normal but also necessary, psychologists say. Lynn Ponton, an adolescent psychiatrist in San Francisco and author of *The Romance of Risk*, says risk-taking by teenagers has been wrongly stigmatized, with parents lumping all risk-taking together into one frightening glob, too horrible to think about. Close your eyes and hold your breath.

But many child psychologists say human teenagers, not unlike Fletcher the monkey, need to do chancy things to find out who they are, where they fit in. And parents need to figure out when the chancy thing is within normal range and when it's moved far beyond—a tricky calculation that can often depend on the kids themselves. For some kids, trying out for the school play or taking an advanced math class is all the risk they need. For others, it's flying over ravines on mountain bikes, walking on the wild side of town, or taking the first gulp of beer. Ponton and others say that teenagers who do experiment in a wide range of areas—even drugs and alcohol in a limited way—often adjust better in the long run than those who completely hold themselves back.

"We used to call risk-taking acting out, and we used to think of all of it as bad," Ponton explained when I went to see her in her office in the hills of San Francisco. "But risk-taking is a normal tool of development. Teenagers define their identity through risk."

As noted earlier, certain damaging behaviors, such as getting pregnant, have, in fact, edged downward in the United States recently, perhaps because of a concerted campaign by grown-ups and schools to deal with the issues. Other risky activities remain frightening frequent, particularly in areas involving violence and sexual disease. Each year, one in four sexually active teenagers contracts a sexually transmitted disease. Nearly half of all new HIV infection cases—about twenty thousand a year—occur in those under twenty-five years of age. Some violent acts by teenagers—knife or fistfights—have declined somewhat, but the number of incidents involving multiple victims—kids shooting one another in cafeterias with semiautomatics—has gone up.

"I would say what's most different for teenagers today is the environment," says Larry Steinberg, a psychologist at Temple University. "The level of stress is taxing their emotional and cognitive resources to an extent not true in the past. Puberty is happening sooner. The drugs are stronger, there's AIDS. And they get less support from adults. Teenagers take their same risk-perception abilities and apply them to situations that are more dangerous and in a context where there is less adult guidance and involvement."

That leaves many modern teenagers at the mercy of their own judgment, or that of their peers. But even peer pressure may not be what we think. Some recent research suggests that teenagers, rather than being pushed by friends, purposely pick friends who do things they want to do. They find out which bus is going where they want to go and hop on. Dan, one young drug-taking teenager I spoke with, told me he'd had all different types of friends at the start of high school, but that, at a certain point, he decided to hang out with the ones who wanted to drink and do drugs, like he did. "I started to ignore the ones who wanted to study and stuff like that," he said.

Another long-held belief is that teenagers do dumb and dangerous things because they think they'll never die or will live, relatively safely, at least through their twenties. They think they're invincible.

But Susan Millstein, a psychologist at the University of California at San Francisco, who has just completed the first long-term study of teenagers' perception of risk, has found they don't necessarily consider themselves immortal, at least not any more than the rest of us. They're scared of all sorts of things: dying, their parents dying, getting hurt, getting bad grades, getting kicked out of their group. Most teenagers don't go around, willy-nilly, putting themselves in harm's way.

Teenagers sometimes don't recognize different options, and can often be forced into making poor decisions in situations that are both emotional and stressful. But contrary to how it often appears, teenagers do think, using whatever skills and knowledge they possess at the time.

"A young girl may truly believe that the only way to keep her boyfriend is to have sex with him; it's a social benefit she's after," says Millstein. They may not be as good at it as adults, but teenagers "weigh costs and benefits just like anyone else."

The Neuroscience of Risk

Over the years, neuroscience has not been a part of the dialogue about risk and teenagers. But that is changing.

Some neuroscientists, like Chuck Nelson, think brain science has already given us some clues to risky behavior in adolescents. After all, one primary tool teenagers use in figuring out whether to climb out a window, have sex, or pop a pill is the prefrontal cortex, that specialized brain part that acts like a policeman and says: "Stop!" If that region is still not fully developed in adolescence, as new research indicates is the case, "that means they just may not see the consequences of their actions," says Nelson.

One scientist who is trying to zero in on risk-taking and decision-making by teenagers is Ron Dahl, a pediatrician and child psychiatric researcher at the University of Pittsburgh Medical Center. Dahl and his colleagues are conducting a long-term, brain-scanning study to find out how or if kids alter the way they make decisions as they progress through puberty—and if so, how. To do this, they are scanning teenagers' brains as they play a computer game that measures how they assess risk. In the game, kids can win more money for a time if they make certain risky choices. But the game is rigged so that the biggest payoff comes if they make more conservative, less-risky choices.

Dahl is attempting to find answers to a number of questions: Will thirteen-year-olds who are far along in puberty use a riskier strategy in the computer game than thirteen-year-olds who have not yet started puberty? If so, what is the cause of the change? Is it the result of hormones or some other neurochemical, or a change in structure in the teenager's brain that makes them more open to risky behavior?

While complete data is not in yet, early results indicate to Dahl that "some aspects of decision-making appear to change with puberty."

As he puts it, puberty is a time when "passions are ignited," not just romantic ones, but also "strong desires to achieve certain kinds of goals." Puberty is associated with an increase in the intensity of certain kinds of feelings. It's a time when, for whatever reason, "emotions are ramped up."

"Teenagers seem to be able to tolerate some fear if something is thrilling," he says. In fact, there seems to be "something about the reward systems that biases their choices and decisions toward excitement, even if there is some risk."

Some scientists think the teenage outpouring of testosterone or estrogen is biasing those choices. While that might be true, Dahl says that, based on his research so far, he believes it's more complicated than simply a hormone effect, perhaps involving complex interactions across several brain systems of motivation and reward, including those that involve the neurotransmitter dopamine—one of the key brain chemicals that carry and influence the messages between nerve cells.

THRILLS, THRILLS, AND MORE THRILLS

Dopamine is well known for its effects in Parkinson's, a progressive disease, most often of the elderly, that causes rigidity and develops when cells in the substantia nigra, a structure deep in the brain, slow their production of dopamine.

So what exactly is dopamine up to in the teenage brain?

The answer is thrills. Aside from helping to smooth motor movements, dopamine is heavily involved in what's known as the pleasure-and-reward circuit in the brain, the part that kicks in when we have something we like, something that makes us feel good. And it is, in part, the good feelings we get from increased levels of dopamine that push us to reach for that same thing again.

Not surprisingly, we've evolved to seek rewards that better our chances of survival. We have sex because we like the way it makes us feel; we eat because we like the taste and the smell of food and the feeling of a full belly. In some instances, many scientists now say, we also take certain risks because through our evolutionary history it, too, has enhanced our chances of survival.

SEVERAL years ago, eight young men in a scientific experiment sat in front of computer screens in a lab in London playing a video game, pretending they were tank commanders navigating their way through a battlefield. If they succeeded in the game, which became harder as they went along, they were rewarded with money.

As the men blasted and dodged their way forward, researchers used a PET scan to measures levels of certain chemicals in their brain by injecting radioactive tracers. They found that one neurotransmitter was enormously busy in a midpart of the brain that's associated with the reward-and-pleasure circuit. That neurotransmitter was dopamine.

Paul Grasby, a psychiatrist at Imperial College in London, one of the investigators, said the radioactive tracer they injected into the young men, most in their early twenties, is known to hook onto dopamine receptors in the brain. When the tracer was unable to find docking places, it was assumed it was because dopamine was already docked there. The study, published in the journal *Nature* in 1998, was one of the first times that dopamine had been found doing its work in a living, reward-and-risk-taking human brain.

There's nothing easy, of course, about sorting out the activity of a single neurotransmitter. In 1940, Grasby says, there were only one or two known neurotransmitters. Today, we have identified dozens.

"We are chasing very small signals and there is always a danger in making a one-to-one correlation with one neurotransmitter and any complex human behaviors such as rock climbing or delinquency," said Grasby. But there is now a common thread of evidence, Grasby agrees, that links dopamine and a "predisposition or a biasing" toward certain risk-taking behaviors.

No one knows the precise levels of dopamine in the brains of teenagers, mostly because it can only be measured by invasive techniques such as PET scans. But because dopamine declines overall between childhood and adulthood, scientists say it's logical that teenage dopamine levels are still considerably higher than levels in most grown-ups. And those higher dopamine levels may, according to Nora Volkow, a neuroscientist at the Brookhaven National Laboratory in New York, leave adolescents much more susceptible to a range of stimulations—and may be one reason why drinking, drug-taking, as well as novelty and risk-seeking begin a steep increase in the teen years. Teenagers may have a built-in urge to get out on the edge—and the means to get there.

Dopamine works in the brain in a number of ways. Volkow, who has studied dopamine for years, says nearly all the addictive drugs, including cocaine, heroin, nicotine, alcohol, amphetamines, and, to a certain extent, marijuana, have been shown in animal studies to increase dopamine in the reward centers in the inner brain, either by jump-starting the initial release of dopamine into synapses or prolonging its stay.

And when the dopamine cells in the pleasure circuits of the brain are activated, in such areas as the nucleus accumbens, for instance, we get, according to Volkow, "a feeling of well-being."

What's more, dopamine is also known to work on areas of the cortex that are crucial for what neuroscientists call "salience"—

that is, our ability to recognize when something is important and take action.

"If you've just eaten and see a piece of ham, that piece of ham will not be important or valuable to you; it will not be salient," explains Volkow.

But if you're hungry that ham takes on new meaning; dopamine washes through the cortex, activating cells, telling you to pay attention, this is relevant, do something. Other researchers describe this action as "error detection." Dopamine increases when we encounter something new, a novelty, something we have to sort out. Is it good or bad, friend or foe?

Dopamine, like all neurotransmitters in the brain, works on a complex feedback loop. If you have a lot of it, you may be prodded to act in a certain way, perhaps take that drug or drive that car too fast or poke your nose where it doesn't belong. And those specific actions, in and of themselves, can, in turn, further increase levels of dopamine. Take one roller-coaster ride and you may find you want another.

But it's also possible to get too much dopamine. Most addictive drugs can produce what Volkow calls a "massive bomb attack" in the brain, overstimulating dopamine neurons. When that happens, the brain tries to blunt the effects by actually cutting back on the number of dopamine receptors, those little molecular baseball gloves on the outside of brain cells that catch dopamine as it whizzes by. It's been shown recently that high levels of stress—a frequent characteristic of the teenage years—can also reduce dopamine receptors. In such instances, the brain may be left with reduced levels of dopamine and an urgent need to get more, perhaps pushing a teenager to take more drugs or push that accelerator down harder.

It's also possible, Volkow says, that some people are simply born with dopamine systems that are either over- or undersensitive, leaving them genetically primed to seek or avoid a wide variety of risky activities.

"I believe that novelty, risk-taking, activates dopamine in humans, no question," says Volkow. "That's also the system nature has of telling the organism to pay attention to the new thing, figure out if it is a positive or a negative. And teenagers, after all, have many more new things to sort out."

As Ron Dahl says, it's important to remember that puberty is "not one thing," but a number of changes in the body and the brain. All the increased emotional intensity and tendency for risk-taking in adolescence may very well be taking place independent of a teenagers' slower development of cognitive function. If that's the case, Dahl says, the fact that puberty is developing two years earlier in girls—and somewhat earlier, he believes in boys, too—than it did a hundred years ago could leave teens today with a "particular vulnerability," and put those who mature especially early at heightened risk. Puberty may rev up thrill-seeking before there is what Dahl calls a "coordination across systems," before the frontal cortex comes on line to say "whoa!"

"Then what you have," Dahl says, "is an engine without a driver."

SEEKING OUT THE NEW

Michael Bardo, a psychologist at the University of Kentucky, has monitored this dopamine connection in rats for some time.

According to Bardo, rats, like humans, are drawn to the new, new thing. A certain level of risk-taking, even in rats, is normal and necessary. Give a rat a choice between entering a small room with old familiar toys and smells, or a room filled with exotic odors and a never-before-tried plastic slide, and the rat, adolescent or not, will almost always choose to poke its pink nose into the unknown.

"We are biologically set up to enjoy new things. That's how we find new sources of food and new partners and find out where dangers are. It's adaptive and it's natural," says Bardo.

To pinpoint where that drive might come from, Bardo gave a few of his rats a drug that blocked dopamine in their brains. He took a few others and surgically removed the dopamine-producing systems in their brains. He then gave the rats the choice: old familiar restaurant or the new one around the corner. Invariably, the dopamine-depleted rats chose the place with the old menu. In other words, the cells deep in the brain that produce dopamine and send it to the reward areas of the brain were not being activated, and the rats were not motivated to do much at all.

In another experiment, Bardo, working with George Rebec of Indiana University, set up a complicated electrode system by which they could measure levels of dopamine in a rat brain. And sure enough, as the rats ventured into new and mysterious places, dopamine levels in their brains shot up, particularly in the nucleus accumbens, one of the reward circuit areas.

All this reinforces the role of dopamine "as a critical neurotransmitter in risk-taking," says Bardo.

Bardo and a number of other scientists believe that the rat—and human—world can be roughly divided between the high rollers and the stick-in-the-muds. Although all rats are attracted to novelty, some—teenagers and grown-ups—almost always choose the new thing, while others get off the couch only occasionally. Bardo believes that such high-risk-takers may be born, not made. Risk-taking varies from individual to individual, he says, like body weight or height.

Not long ago, a short time after John F. Kennedy Jr. took off in his plane in twilight haze, without a flight plan and with a bad leg, and crashed into the ocean off Martha's Vineyard, a newswire story carried the headline: "Kennedy Tragedies Linked to 'Risk-taking Gene,' Israeli expert."

That expert, Richard Ebstein, director of research at Jerusalem's Herzog Hospital, was the head of one of two teams of researchers who several years earlier, in January 1996, reported that novelty seekers tend to have a particular type of a gene that allows the brain to respond to dopamine. The risk-takers, the researchers said, had a variant of the dopamine D4 receptor gene that was slightly longer than the gene for that receptor in more timid people. The theory, outlined in a front-page story in the *New York Times*, was that the long gene "generates a comparatively long receptor protein," and that somehow that oversized receptor influences how the brain reacts to dopamine. People with the long gene type were significantly more likely to describe themselves in questionnaires as "impulsive, quick-tempered, fickle, curious, and extravagant" than were those with the shorter gene.

Several months later, however, another study—this one based on research in the United States and Finland—seemed to contradict the earlier findings. These researchers found no link between thrill-seeking behavior and the longer, D4 receptor gene.

So where does that leave us?

Richard Ebstein says there's no question in his mind that novelty seeking has a genetic basis. Studies of identical twins through the years have found that a trait like height is about 80 to 90 percent inherited (with diet often making up the difference). And the same studies have consistently found that a novelty-seeking personality is about 50 percent heritable. That means identical twins have a fairly high chance of having the same risk-taking profile even if one is raised in Soho and the other in Sioux City. But today, Ebstein is more doubtful about finding any one gene to blame. He recently completed what's called a meta-analysis, a review of all the recent studies of the D4 receptor, and the findings, he admitted, were all over the map. Some studies found a link and others did not.

The case illustrates how difficult gene hunting can be. When scientists have tried to buttonhole a single gene for a highly complex behavior—say, the need to bungee jump—they have almost always failed.

But Ebstein, for one, doesn't feel we should give up. The trait for novelty seeking, he says, is an outgrowth of one of our most basic instincts, approach or avoidance. That trait is not only easy to spot on questionnaires, but it's an integral part of every living thing, up and down the food chain. It's fairly easy, for instance, to spot avoidance.

"Even if you put a little vinegar in water, the paramecium will all rush away from it," Ebstein says. He reminded me of the now-famous studies by Harvard psychologist Jerome Kagan, who found the same thing in young children: A certain percentage of young children are attracted to novelty, such as new visitors to their kindergarten class, but another group shies away from such changes. Such fundamental traits, Ebstein says, are likely to have a genetic basis, and he believes that genetic basis will turn out to involve dopamine. More inhibited behavior, he says, is likely to be linked to another more calming neurotransmitter, serotonin.

But Ebstein acknowledges that, even if a genetic basis were confirmed, it would be only part of the story. Some of the Kennedy family, for example, may have a propensity for higher-risk behavior, Ebstein points out, but they are also "good politicians and the life of the party." It's not so much what gene you have, he says, but what you do with it. A novelty seeker in one environment might join the Special Forces and hunt down terrorists; another might become an alcoholic.

"One is socially acceptable and one is not," Ebstein says.

In a large basement lab in Binghamton, New York, researcher Linda Spear has been running dozens of experiments with adolescent rats. As chairman of the psychology department at the State University of New York at Binghamton, Spear has devoted a great deal of her career to figuring out why adolescent rats do the strange things they do.

One late fall morning I went to Spear's lab to watch adolescent rats act just as weird as their human counterparts. In one experiment, a lab aide carefully carried one white rat to a small room with a wooden structure that contained two connected covered runways and two uncovered runways, all about a meter off the floor. Spear explained that all rats are afraid of heights, and adolescent rats generally don't like to be on the elevated maze at all. As we watched the rat's movements on a small TV monitor in a room next door, the adolescent female immediately went out on the open runway. But she did not stay. After a second or two, she turned back and spent most of the rest of her time in the closed areas, poking her pink nose up hard against the walls.

In fact, Spear says, adolescent rats are sometimes less likely to take risks than adults. Adult rats go out on the open arm runway much more readily.

Still, sometimes the opposite happens. In a second experiment at Spear's lab, a graduate student plopped another rat into a small plastic swimming pool filled with sand. Generally, Spear explains, adolescent rats are not at all afraid in the sandbox. In that situation, she says, adolescents will "explore much more than adults." And sure enough, the rat scurried resolutely to where a Froot Loop was buried, without a rat care in the world.

Spear says this odd mixture of greater willingness to take risks, but at the same time, in truly frightening situations, more caution, is typical of adolescent rats. If the activity is too scary, she says, adolescents are more vigilant. But if the situation is moderately risky, they will take many more chances than grown-ups. (Spear and her colleagues have also shown that adolescent rats are consistently more social and seek out novel objects more than grown-up or kid rats.)

Spear believes that the only way to explain this contradiction is to look at evolution.

"All mammals are faced with the same problem," she told me over lunch. "They have to go from dependence to independence. But they also have to make sure they don't mate with their parents in the process."

To ensure that doesn't happen, in most species one gender leaves the group at adolescence and joins another group for mating. And to push themselves to do that, she says, some attraction to risk is necessary. Rats in the wild stick their noses out of their burrows and start to scavenge only when they hit puberty. But it's scary out there. Young adolescents aren't that good at getting food; they aren't very strong and are vulnerable to predators. If they are not careful, they die.

"So what's good for the species is not necessarily so good for the individual adolescent," said Spear. Adolescent rats—and humans to the extent that such behavior carries over to us—

have to have an odd combination of "increased risk-taking and vigilance."

Spear first started thinking about adolescence in the 1980s when she noticed that most of the developmental curves she was plotting—for the animals' responses to drugs or stress or alcohol—got horribly balled up when they reached adolescence. Her curves would show a nice trajectory and then, as a rat hit its teens, start "wiggling all over the place," leaving Spear frustrated and annoyed. "I got interested because those adolescent animals really began to tick me off," she says.

Spurred by that frustration, Spear recently wrote one of the most comprehensive reviews of emerging science of the teenage brain. And she found that there were indeed dozens of ways that teenage brains—in rats, monkeys, and humans—were different from younger or older brains.

"[The] remodeling of the brain is seen in adolescents of a variety of species," she wrote in her review. "[And] the focus of research is gradually changing, with the recognition that the brain of the adolescent differs markedly from the younger or adult brain, and that some of these differences are found in neural regions implicated in the typical behavioral characteristics of the adolescent."

And one of the big ways Spear believes that adolescent brains differ is in connection to dopamine. During adolescence, the level of dopamine is generally declining from peak levels in childhood. But, at the same time, it is still increasing in at least one key area of the brain—the prefrontal cortex. As it increases in that late-developing region, enabling the brain to establish the connections it needs for a lifetime, Spear says, the brain, seeking to retain a balance, correspondently decreases dopamine levels in the nucleus accumbens and the rest of the inner reward circuit of the brain.

What might a drop in the brain's reward areas mean in terms of adolescent behavior? Spear believes it may mean a great deal, that teenagers as a group, perhaps dopamine-depleted in their reward system, might need more stimulating activities to get the same "kick" as we get. "They may need more to get the same bang for their buck," she says.

B. J. Casey, an expert on scanning children's brains, agrees there may be a connection between dopamine and adolescence, but she looks at its actions in the frontal cortex. Since dopamine helps us to notice things and take action more quickly, it might have a special job at that time in life in the prefrontal cortex. As it seeps into the frontal lobes, it may help teenagers recognize the "new" thing, decide it's important or "salient," and act quickly.

"Maybe in evolution it was important to have some trigger, something to help us grab the moment, maybe go after that mate," she says.

Looking for the Edge

Several years ago, in an experiment in a middle-class high school, Michael Bardo and his colleagues divided a group of students into high-risk-takers and low-risk-takers based on their answers to a questionnaire, such as "I would like to jump out of an airplane," or "Dangerous things scare me." Once the group was divided, the researchers followed them throughout their high school years. In analyzing the results, they found that the highest-risk-takers were ten times more likely to take drugs, which experts say in our culture has become one of the most common forms of risk-taking by teenagers.

Over the past several years, Scott Lane, a neuropharmacologist at the University of Texas at Houston, has studied a group of thirty teenagers, both boys and girls, who are considered high-risk-takers. These are the kids who have already been arrested, who started using drugs at age ten or twelve and didn't stop, who've run away, who've dropped out of school.

In his lab, Lane and his colleagues set up a game that resembled gambling or, in certain ways, life, a version of a typical lab experiment that is often used to study decision-making. The teenagers were told that the goal of the game was to earn as much money as possible. As they played, they were given a choice between two buttons. One of the choices was set up to be safer, and pay out a small amount, or in certain situations require that they give back a bit of money. The other choice paid out seventy-five cents—a lot of money in the game—but also demanded that they pay back more money in other situations. The game was rigged so that the safer, more predictable choice would, in the long run, pay off considerably more.

When Lane compared the results of the high-risk teens with a group of kids whose risk-taking was in a more normal range, the results, in more than a thousand trials, were conclusive: The high-risk-takers chose riskier strategies and lost the most money. They were, as Lane described them, "oversensitive to reward." If they won, they were motivated to keep pushing the same button four or five times more, even if the strategy no longer worked. The control group, on the other hand, would more quickly "drift back" to a safer strategy and would come out ahead in the end.

Lane says he believes "this higher sensitivity to reward—and excessive risk taking—is clearly related to dopamine." Studies by Lane and his colleagues show that low-risk-takers can be turned into high-risk-takers if they're under the influence of alcohol which, among other effects, is known to increase levels of dopamine. (Linda Spear finds that adolescent rats stay out on the scary open arm maze only after they've had alcohol.)

Although environment clearly plays a key role in determining any behavior of humans, Lane's study had kids from the same families in both high- and low-risk-taking groups, another clue, he said, that something might be awry biologically. "There might very well be a dysregulation in the dopamine system in extreme high-risk-takers," he says.

Still, he is not sure whether the dysregulation comes from too much dopamine or too little; the brain would try to compensate for either circumstance. What we do know is that adolescents in general are higher-risk-takers than adults, and that something makes some teenagers take bigger risks than others, something makes them push that accelerator down a bit more.

"It is very hard to implicate one neurochemical in any complex behavior; in fact, it's a gross oversimplification," acknowledges Lane. "But my guess is that in the end dopamine will be at least one piece of the jigsaw puzzle."

BY UNDERSTANDING how a normal brain works, scientists hope they can eventually find ways to protect certain teenagers from their own, extremely risky behavior and destructiveness.

In one such effort, Tom Kelly, a professor of behavioral science at the University of Kentucky, gathered together a group of young men and women who, in earlier studies, had been identified as high-risk or high-sensation seekers in their teens. In the ongoing study, Kelly is giving those high-risk-takers, now all over eighteen, doses of amphetamines and then having them do such things as watch the movie *Speed*, with its hair-raising bus-driving scenes, or play video games with frantic motorcycle chases. What Kelly's trying to find out is whether one thrill can be substituted for another, whether the high that comes from watching a thrilling movie scene can somehow blunt or replace the impact of the amphetamine. After such activities, Kelly measures how subjects do on a battery of motor and verbal tests. The question he is asking is: Will high-risk-takers do better on tests, be less influenced by amphetamines, if they are already stimulated by a safer alternative?

Kelly hopes such studies will help identify those kids whose risk-taking is likely to be destructive. In an earlier study by Kelly and his colleagues, high-sensation-seeking teens were found, when they became young adults, to be much more sensitive to drugs such as amphetamines, and more likely to say they liked them. In other words, says Kelly, "there may be kids who are more vulnerable [to the drugs] and those may be the kids we need to spend more time working with." It would be nice, he says, if researchers could pinpoint a safe thrill that could effectively replace the unsafe one.

The point, adds Bardo, is that we have to recognize that kids take risks and realize that some are high-risk-takers and then "get them to take risks in a more pro-social way."

NARROWING THE CHOICES

A great many educators and parents—and even some kids themselves—worry that in American culture we've eliminated too many of those positive risks for teenagers—hardly anybody builds and sails their own rafts down rivers anymore—leaving teenagers with only a few limited ways to be adventuresome, most of them laced with sex and drugs, and illegal. "Modern rebels are lazy," one sixteen-year-old girl told me. "The risk-taking these days is only about drugs."

My husband, Richard, grew up in a small, rural New England town of five thousand. The "risks" he took seem relatively quaint by today's standards. He and a few friends would drag race old beat-up cars, doing "quarter milers" on deserted country roads. That was dangerous. But since there were few others on the road, the danger was fairly mild. He also spent hours roaming alone with his dog through the woods, deep in the middle of nowhere. No one knew where he was; the decisions he made were his own. Sometimes, with his small-gauge rifle, he'd shoot a squirrel.

I grew up in the racier environment of a Southern California beach town. By the time I was a teenager, my high school, strong in academics, also was awash in marijuana, speed, and sex.

Even so, there seemed to be more options and we had thrilling things to do that didn't involve pot or vodka. In fact, we spent most of our early adolescence taking as many risks as we could on the giant waves of the Pacific. We weren't always studying for tests, we were at the beach. Our parents had only vague notions of where we were or what we were doing. We spent hours shooting our bodies through enormous ten-foot waves, waves so big I look at them now and shudder.

MANY fear there are too few outlets for such normal risk-taking today—and too limited paths toward success generally. One longtime high school guidance counselor in a high-end suburb of Washington, D.C., trying to explain the increase in violence and drugs in schools, told me recently that she knew why so many kids seem to get in more serious trouble today. "There's no question in my mind," she said. "We've simply made schools impossible for the regular kid. There are not enough options for how to be a successful teenager."

It's an idea that teenagers themselves, I found, often give voice to. One sixteen-year-old girl said she thought kids turned to drugs, in part, because they found the world was too black or white, with academic success being the only barometer by which to measure what it means to be successful. Drugs were an easy way out of a seemingly impossible game.

"It seems like you have to go to Harvard—or you will be a druggie and a dropout," she said. "There don't seem to be any in-between choices. People just talk about getting into a good college all the time; they pound that into you. They never talk about being a nice person or having a good marriage or a nice family. It's all about grades. And there isn't any room for mistakes."

Jay Giedd at the National Institutes of Health agrees. In his private practice as a child psychiatrist, he sees a number of high-end, academically talented kids get into emotional trouble because, he says, they only see one road, one way to be a teenager.

He told me about one sixteen-year-old girl who didn't make the soccer team and feared that would doom her chances of getting into a good college. She saw no options, he said. With one setback, she felt like a "complete failure" and started having panic attacks in the bathroom. Another boy, seventeen, a gifted student, had a life that is not unusual for fast-track kids today. He went to school, played lacrosse, and did homework, all so he could get into a top competitive college. He almost never had free time to just go outside and mess around, and refused to let himself take minor risks or make small mistakes. Then, one year, although he was still one of the top students at his school, he found he was no longer the top student. He began to believe his life was over. He froze. He could no longer do even simple homework assignments and he fell into a deep depression. His parents, Giedd says, were well-meaning, but they were mostly concerned about "how long the depression would last" and whether it would hurt their son's chances of getting into a good school.

"I see kids who are cracking up because of the stress of the workload and because they see only one way to success, to getting a good job," said Giedd. "They don't take many real risks because they are afraid. But, maybe because of that, they have not learned to make their own decisions. That worries me. I think kids need to learn life's lessons, somehow, early. They need to take risks, to make some mistakes."

Chapter 8

Corny Jokes and Cognition

The Adolescent Brain Starts to Get It

Here's my favorite stupid joke:

Two old guys are sitting on a bench and one says to the other: "Hey, I got a new hearing aid and it's the top of the line, works great."

"Wonderful," says the other guy. "What kind is it?"

"Oh, about five-thirty," the first guy says.

It's a bad joke, I know. Still, I bring it up because it, too, has a connection to how a teenager's brain grows.

Along the way, there are moments that crystallize in the minds of parents. The first grin at peek-a-boo, the first successfully buttoned jacket, the first meticulously drawn lollipop tree. Like any parent, I thought those moments were thrilling. Still, for some reason, one of the most thrilling moments came later. It was when my budding adolescents got subtle, ironic—and corny—jokes.

For years, my husband and I would tell stupid and no doubt overly nuanced jokes, largely to a mute audience. Then, one day, something changed. One stupid joke about a hearing aid and, instead of a blank stare, there was, sometime near the first stirrings of early puberty, a slight glimmer of recognition, a faint light in the eyes, a distinct and audible: "Ha!"

To me, it was one of the true—and in this case, welcome—markers of emerging adolescence, evidence of a shared view of the world, a cognitive, albeit corny, connection with the grown-up world.

Most parents, on some level, are well aware that something changes—and changes a lot—in the average, normal ability of the average, normal teenager not only to get the irony of a joke but to think abstractly or understand the emotions of others. It's a process that's both familiar and largely taken for granted with, as one mother told me, "very little thought as to where it might come from."

The mother of three teenage boys told me they seemed to get a "broader view" of themselves and the world somewhere around age sixteen. "My twelve-year-old still has a very narrow ability to view the world; he sees things as one dimensional," she said. "But the sixteen-year-old has this capacity to view things in a larger way. He no longer thinks he's God's gift to the world, for instance. He has, you could say, a wider definition of what's cool."

Another mother was struck by her daughter's growing concern for others. Her daughter's Latin teacher had given up her Saturdays to help a group of high school students prepare for an exam, and the kids got together on their own and bought the teacher a big present. "It may seem like a small thing but I don't think it would have occurred to these kids even a few years ago; they really seemed to understand that this person had sacrificed for them and they wanted to give her something back. It was altruistic in a way; it was the ability to think about the needs of others." A mother of four, ranging in age from thirteen to twenty-seven, said she found they generally started to "explode intellectually" around age sixteen.

Often the teenagers themselves recognize a shift.

"I remember walking by a building and I looked at its 'For Rent' sign and, for the first time, I had all these weird thoughts," said Sarah, sixteen. "I didn't just think of it as a building, but I thought, gee, someone had to paint that sign and someone had to make that building—probably there were dozens of people who worked on the building to get it there. And, to me, it was the first time that I really realized what a big world it was and that it has a lot of people in it and I began to think: Where do I fit in?"

Eric, seventeen, on his way to Brown University, told me he found that over the past year or so he could feel his brain "think more logically," with many more "abstract kinds of thoughts," and an ability to see underlying relationships.

"When you think about things like a neuron in the brain, for instance, it's very small and you can just think about that, but then you can also think about how out of all those small neurons come big things like anger and happiness," he said. "I've been thinking a lot lately about how big things are made up of a lot of smaller things. I seem to be applying that one thought to lots of areas and I don't remember doing that before. I can see connections now and I like making those connections. Sometimes, now, I find it just feels good to think."

Ian, also seventeen, said his mind now makes leaps it never made before. Highly experienced in the martial arts, Ian said that now when he learns a new technique to defend a certain punch, his mind instantly sees how that same defense could be used with different sorts of punches as well. While a few years ago, he says, he might have described himself as "all limbs," now he feels that his physical and mental senses are working together, and he can do gymnastics in the air with both. "It's like when you realize you don't have to write down a math problem, but you can do it in your head," he says.

Where does it come from? Does a kid suddenly see invisible connections in the world after so many days in geometry class? Do they get a silly joke after some preset number of hours watching *Friends* on TV? Does some synapse somewhere deep in a remote valley of the cortex suddenly sit up and say, Ha!

Over the past thirty years, academic researchers have tried mightily to pin down the precise path of a teenager's cognitive and emotional development. How and when does a teenager see not only the subtle irony in a joke but the nuances behind the concept of justice, the shades of gray behind a "maybe"? And, equally important, how does a teenager use those new cognitive thought processes to make decisions about risks, life, everything. Should I get in that car? Should I be his friend? Should I go to that wild party? Who am I?

Kurt Fischer, a psychologist at Harvard, has, over the past few years, developed an outline for how he believes cognitive and emotional development proceeds in teenagers. It follows a course, he says, that largely mirrors the physical growth of the brain.

Fischer and several other researchers, using mostly electroencephalograms, or EEG measurements, of electrical energy generated on the outside of the skulls of children of various ages, say there are clear "growth spurts" of higher neural energy at certain times in children across cultures. Fischer believes these spurts, which he says happen roughly at ages four, eight, and eleven weeks; four, eight, and twelve months; and two, four, seven, eleven, fifteen, and nineteen years, "coincide with cognitive development." The spurts happen, Fisher believes, when humans are acquiring certain skills—when a baby can remember where a toy is hidden, or perhaps when a teenager gets a joke.

In some ways, such ideas align with those of Jean Piaget, the Swiss child psychologist who concluded that intellectual and emotional abilities develop in stages. At a certain age, around five or six, for instance, most children will begin to understand that a short, fat glass could have the same amount of liquid as a tall, thin one, a concept they miss earlier. As children move toward adolescence, Piaget believed they begin to think more abstractly—they are able to deal with ideas such as justice and honesty, able to do backward logic, get a pun, see hidden irony.

As Fischer outlined this process in his book *Human Behavior and the Developing Brain*, a child at ten to twelve would understand honesty as a "general quality of being truthful." But by age fourteen to sixteen, they would have acquired more abstract, less black-and-white reasoning, capable, for example, of seeing the value in a social lie. While a younger teen might see a parent as a hypocrite if he holds two opposing views, an older teenager would begin to understand how two things can be true at the same time, and weigh the evidence for each.

Fischer, who takes this idea further than most, is convinced that the stages of cognitive and emotional growth, if a human is exposed to normal environments, are tied to spurts, or a "reorganization of neural circuits" in the brain. He was not surprised at all by Jay Giedd's findings of a growth spurt of gray matter at puberty. He thinks neuroscientists will find many more down the road.

"There's a straightforward relationship between brain growth and cognitive and emotional growth," he says.

One way we are learning more about what teenagers know has been forced upon us. After a flurry of school shootings several years ago, the country was confused. Here were: High school students Eric Harris and Dylan Klebold, who shot to death twelve fellow students and a teacher at Columbine High School; Nathaniel Brazill, who at age thirteen shot his favorite teacher; Michael Carneal, who at age fourteen killed three students; Mitchell Johnson, who at age thirteen, along with Andrew Golden, who was only eleven, killed a teacher and four students; Kip Kinkel, who at age fifteen killed his parents and two students. What to make of these kids? Where was their cognitive and emotional growth? And what were we to do with them? Out of the confusion came a hodgepodge of actions. Some states, like Florida and Kentucky, tried them as adults; others decided they were juveniles, not fully responsible for their actions.

So what do teenagers feel and when do they feel it? What do their brains know and when do they know it? When do they know that their actions might have horrible outcomes? Certainly, in the extreme cases of the teen shooters, the crimes were adult-sized, but does that mean those teenagers are adults in every thinking and feeling way? When exactly does a child know that if you shoot that gun at someone they will be dead, forever?

In March 2001, after baby-faced Charles Andy Williams, slight and smiling, opened fire in his suburban California high school with a .22-caliber pistol, killing two fellow students and wounding thirteen other people, Daniel R. Weinberger, director of the Clinical Brain Disorders Laboratory at the National Institutes of Health, wrote an extraordinary article for the op-ed page of the *New York Times*. His op-ed piece pointed back at the frontal lobes. He wrote:

"To understand what goes wrong in teenagers who fire guns, you have to understand something about the biology of the teenage brain, and the brain of a fifteen-year-old is not mature, particularly in an area called the prefrontal cortex, which is critical to good judgment and the suppression of impulse control.

"Everyone gets angry; everyone has felt a desire for vengeance. The capacity to control impulses that arise from these feelings is the function of the prefrontal cortex . . . it takes many years for the necessary biological processes to hone a prefrontal cortex into an effective efficient executive. The fifteen-year-old brain does not have the biological machinery to inhibit impulses in the service of long-range planning.

"[This] is not meant to absolve criminal behavior or make the horrors any less unconscionable. But the shooter at Santana High, like other adolescents, needed people or institutions to prevent him from being in a potentially deadly situation where his immature brain was left to his own devises."

Perhaps predictably, the article elicited a rash of responses, some of which pointed out that there are millions of fifteen-year-olds out there who go through whole days without shooting anyone.

"Until recent times, children who reached biological maturity, typically around the age of thirteen, were treated as adults," said one letter from a professor of psychiatry. "Benjamin Franklin was an apprentice printer at age twelve and his brain was evidently well enough developed to plan for the future . . . such fashionable neurologizing of bad behavior is destructive of civilized discourse and human relations."

In the end, whether you can blame the prefrontal cortex and its lack of maturity for such things as school shootings, I really have no idea either. Drawing a straight line from new brain science to complex human behavior is tricky, and even Weinberger, in his essay, stresses that culture, circumstance, violent entertainment, lack of accountability for deviant behavior, broken homes, all "may play a role in the tragedies of school shootings."

What Do They Know—and When?

Clearly, the question of what teenagers know, and at what age, is unresolved. Still, policymakers, concerned about recent violent incidents as well as a host of other questions, such as when a teenager should be permitted to get an abortion without parental consent, continue to wrestle with the question.

One researcher who has been trying to sort this out is psychologist Larry Steinberg. As head of the MacArthur Foundation's Network on Adolescent Development and Juvenile Justice, he is charged with developing guidelines to offer courts, to tell them what we can know about what teenagers know and when they know it, to tell them whether, as Steinberg says, someone like thirteen-year-old Nathaniel Brazill "could have stopped himself" before he shot his teacher.

To try to figure that out, Steinberg, along with Elizabeth Cauffman, a researcher and psychologist at the University of Pittsburgh Medical Center, have most recently studied twelve hundred teenagers who have committed serious felonies in Philadelphia and Phoenix, to see if they can, using a theoretical model, define intellectual maturity. They are trying to pinpoint the moment when a teenager, on average, can make a reasoned decision.

And when does that magical moment happen? Although final data are not yet in, Cauffman said the researchers were likely to "draw the line of maturity" somewhere between ages sixteen and seventeen.

"I think the shift really starts at seventeen," Cauffman told me. "That's when we really start seeing a difference."

To define maturity, the researchers, in part, used a model developed in earlier studies, including one used with a group of eight hundred "normal" kids in middle-class high schools in New Jersey and Pennsylvania. That sample was split evenly between boys and girls and had a wide ethnic mix; it was then compared with a similarly mixed control group of two hundred adults.

Both the teenagers and adults answered questions aimed at determining thinking and behavior in three specific areas: responsibility—that is, the degree to which they felt they could depend on themselves, be self-reliant, and resist peers; perspective—the ability to think about consequences of actions ahead of time, as well as the possible impact of actions on others; and temperance—a measure of impulsivity and the ability to regulate emotions.

To sort out those qualities, the researchers used standard psychological questions that had, in previous tests, been repeatedly linked to certain behaviors. The survey, based on self-reports, asked participants to rate themselves on such questions and statements as: "Do you think that hard work is never fun?" "Do you think that, if you are not the leader, you should not suggest how things should be done?" "If something more interesting comes along I usually stop work I am doing," or "I will give up happiness now to get what I want in the future."

"We found that, as a group, adolescents are less mature and have less perspective and less impulse control than adults," says Cauffman. "Now, I realize that anyone who has an adolescent would say they don't need a big study to show this, but when we are dealing with making policy, we have to show it."

The study, Cauffman said, found that the biggest shift toward mature thinking occurred between eighth and tenth grades. It also found that, on average, girls matured much faster than boys, on a trajectory that remained steady from the eighth to the eleventh grade.

"But," Cauffman quickly added, "between the ages of eighteen and twenty-four, the boys catch up. My husband wants me to be sure and point that out when I'm talking about this. The gap goes away; girls can't feel superior forever."

Within all this, of course, there's wide variation. Some boys mature faster than girls. Some kids worry and think ahead from the get go. But, again when considering any policy implications, Cauffman says, researchers must focus on broad trends: "We have to stick to the aggregate." It's also true, as always, that a whole shopping list of ingredients play into maturity: parenting, environment, temperament. As Cauffman says: "No one factor can predict these things."

Still, she too, firmly believes there is, along with those ingredients, a distinct biological process in the brain that contributes to changes in the level of maturity, and new findings by neuroscientists in that area have left her encouraged. The new brain science, she says, "is confirming what we are finding; they are seeing it biologically. To me, it's very exciting to find that measures of such things as regulation of impulse control go along with development of the frontal lobe, for instance."

As someone who has spent her career studying youths in jail, Cauffman is not shy about saying that she hopes the new research, in all areas, will have an impact on public policy, in particular the often heartbreakingly difficult decisions about whether to try young teenagers who have committed violent acts as adults or juveniles. While no one is saying that teenagers are incapable of making sound decisions, Cauffman says that the research confirms, at least for her, the idea that teenagers and adults are, in a range of areas related to brain development and decision making, quite different.

"There was a reason years ago that they set up juvenile courts," she says.

Steinberg, too, says there's little question in his mind that, while development of a human being is extraordinarily complex, basic biology, including the development of the brain, is part of the adolescent picture.

"Where do kids get the ability to see consequences?" he asks. "I think there's a point where kids are not distinguishable from adults, and when they are younger they clearly are. I'm confident there is a developmental curve. It's probably a combination of having enough experience—making enough mistakes when you didn't think ahead so that you realize that that's important—and brain development—having enough computing power to think ahead. And that probably has something to do with development of the frontal lobe, the part of the brain that's devoted to planning and executing tasks."

SHIFTS IN COGNITION

Dan Keating of the University of Toronto, who recently reviewed the scientific literature of cognitive development in normal teenagers, says that, while disagreements continue at far ends of the spectrum of thought, there's a growing consensus that something is going on involving the development of different forms of thinking "between preadolescence and midadolescence."

There's clearly a change, for instance, in what might be called reverse logic: If I say when it rains the lawn gets wet, it obviously does not mean that every time the lawn is wet it's because of rain. A younger teen might not get that. They're not that good, Keating says, at coming up with alternate hypotheses for things that occur. Their minds, instead, work more along concrete and simplistic lines: She frowned at me, she must hate me. If you have anthrax attacks and you have an existing conspiracy, then the anthrax must be part of that existing conspiracy.

But as teenagers age, they start to challenge outcomes, to think through all the implications; they start thinking a bit scientifically. Maybe it was a hose that made the lawn wet. Maybe she frowned because she was sick or having a bad-hair day. With the recognition of alternate outcomes might very well come the ability to get irony in a stupid joke, Keating says.

an inner fenced area. The center is on a dirt road with long stretches of grassland in the beautiful back-river country of the Potomac, about as close to an African savannah as one could get in a short drive from Washington, D.C. And for the past seventeen years Suomi, a fast-talking man with a dry sense of humor and a passion for monkeys, has overseen a rhesus colony here, a four-generation group of seventy monkeys. Not far from where Suomi and I stood was a jungle gym where three young female monkeys lounged in the sun. Off to one side was a small lake with a short bridge to an island, constructed after teenage monkeys continued to swim there, getting scared and stuck. "We had to keep getting out the NIH rowboat and rescuing them," said Suomi.

As Suomi spoke, a thin, gray monkey strutted by. She stopped, looked up, then sauntered on.

"Oh, that's Molly," Suomi said. "She's twenty-six, the oldest, the dominant female. She runs things, settles squabbles."

Rhesus monkeys live in strict female-dominated societies. As Suomi explained, any monkey directly related to Molly would have higher social status as a result, but when Molly dies, "the family drops in dominance."

Known as one of the smartest monkeys—they can play simple video games with joysticks—they are also among the most successful primates. Indigenous to India, they've found ways to live in rain forests or savannahs or on the edge of deserts. Rhesus monkeys are Old World monkeys, meaning they came from Africa or Asia and are more similar to humans than New World monkeys from South America. Some Old World monkeys—apes and gorillas—share 98 percent of human DNA. Chimps share about 99 percent. Rhesus monkeys share about 95 percent of our genes.

And so I asked Suomi, how do these very similar-to-us monkeys experience adolescence? Do they exhibit any distinct behaviors we might, without being overly anthropomorphic, label as teenager?

"Oh yes," he said, "and it's very, very interesting." According to Suomi, rhesus monkeys follow a teenage track much like humans, although they race through it faster. Because they have a similar

Another shift occurs in concepts of self and society. If a teacher appears to be acting unfairly, a younger teenager will see it as unfair and that's it. An older teenager, Keating says, can wrestle with the idea of fairness. How fair was it? Who needs to make some alterations so that it's fairer? Who should apologize first?

As teenagers develop more intellectual skills, however, there can be drawbacks. One exceptionally nice boy I know, who recently got perfect SAT scores, nevertheless went through a period of lying in his early teens that, before it passed, made his parents crazy. "I was really concerned," his mother told me. "I thought, 'Oh, no, he's gotten all my bad genes.' " Another boy, on his way to a good college, said he went through a time in his early to midteens when he stole small things from stores. Why? Because it seemed an "easy and cheap" way of solving a problem of getting what he wanted, he said.

Chuck Nelson, the neuroscientist at the University of Minnesota, said that at a certain point, many teenagers start a rash of lying, convincingly telling parents "they stayed at Susie's house overnight when they really didn't," for instance. As far as Nelson is concerned, this, too, can be viewed as a way of solving a problem, and could again point back to the development of the frontal lobes.

"Above and beyond all the psychological aspects—the need they have to separate from you and all that—lying can be seen a way of getting rid of a lot of problems," he said. "They think, hey, if I just tell them this, I will get away with what I want to do. It's future-oriented thinking, it's a bit like chess—and that's the frontal lobe."

Still, even with his well-grounded belief in the importance of frontal-lobe development in teenagers, Nelson concedes that, as a parent, a sudden burst of prevarication in one's offspring can be unsettling at best. Not long ago, Nelson found three empty beer cans in the trunk of the car that had been driven by his otherwise wonderful sixteen-year-old son. And when Nelson and his wife confronted him, his son glibly told a whopping lie. He told his parents that the mother of a friend had asked him to take the beer bottles to the trash and he had "accidentally" put the bottles in the trunk.

"At first I even believed him," said Nelson. "But then my wife—she's a prosecutor—took me aside and said, 'Are you nuts?!' "

THERE remains plenty of debate in the field of cognitive development in adolescence.

Deanna Kuhn, an expert in cognitive development at Columbia University, for one, believes that by the time a brain reaches adolescence, its development is dictated much more by external influences than internal growth spurts. What matters most at that point, she says, is what you pour into the brain. Whether someone reaches a "higher order" of thinking mostly "depends on the experiences they've had and the culture they're in." Elliot Turiel, a developmental psychologist at the University of California at Berkeley, who has long studied moral development, says that although there's an undeniable "shift" in moral reasoning as children become adolescents—one that involves thinking on a grander scale than one's own immediate self-interest, or the ability to see possibilities for compromise in competing rights or wrongs—much of that comes from moral lessons learned through experiences in the world.

But recent work by neuroscientists continues to point at biology's role.

For example, while many have considered human cooperation primarily a learned behavior, there was evidence recently that it may also have biological roots. Scientists at Emory University in Atlanta, in a brain-scanning experiment, found that when humans cooperated with each other, their brains lit up in the same neighborhoods that come to life when we win a prize, eat a piece of chocolate cake, or sniff cocaine—the inner reward circuitry that responds to dopamine and provides that glow of pleasure. In other words, we cooperate because it makes us feel good.

And that may mean, some researchers speculate, that the urge to cooperate is, at some level, simply innate in humans. Perhaps our early ancestors needed to help each other to hunt big game, find more nutritious food, or raise smarter kids. The ones who successfully learned to work as a team might have had a survival advantage.

Dan Keating who has looked at all the science on adolescent cognition to date, says that, in his opinion, there is little question that there are profound and universal shifts in how teenagers view the world and themselves. It's also clear, he says, that these shifts occur precisely when their bodies and brains are also changing. Keating doesn't believe we have enough evidence yet to point to one specific brain or body event to explain changes and growth in teenage thinking. Still, because the shifts at adolescence appear to involve, he says, such a "dramatic reorganization of social and emotional and cognitive" processes, he, too, thinks the frontal lobes are key. It is, after all, the job of the frontal lobes to integrate social, emotional, and intellectual ramifications of things, to see context: He dissed me. Has he dissed me before? Do I care enough about him to diss him back?

"I think in the end we're going to find that the frontal lobe is where the action is," he says.

Locating the Moral Compass

Antonio Damasio, one of the country's most respected neuroscientists, and his colleagues at the University of Iowa recently reported that they had found, located in the frontal lobes, evidence of the possible biological roots of a human's moral sense.

Damasio described two young adults, a woman twenty and a man twenty-three, who had suffered early injuries to their prefrontal cortices. The man had a brain tumor removed when he was three months old and the woman had been run over by a car at fifteen months. Initially, they seemed to recover well. Both were highly intelligent and grew up in stable homes with college-educated parents; their siblings were fine. But as they grew up, problems emerged. Both became disruptive and incapable of planning and making decisions. As teenagers, both lied and stole, had no sense of right or wrong, and, while often pleasant, showed no remorse for their actions. In Damasio's view, they had not developed a moral compass.

"If you have adults with this kind of damage, they have already learned the social conventions," Damasio told me. "They may apply them poorly in real situations, but they know the rules."

Other scientists say it's important to remember that the subjects in Damasio's study were both severely injured, and that children with less drastic head injuries in similar parts of the brain often go on to develop perfectly normally. But at least with the two seriously injured children in his small study, Damasio says it seems to him that they didn't "learn the rules to begin with." Damasio thinks that's because the prefrontal cortex is one of the main brain areas that learns from punishment and reward, from "pain and suffering." It learns what is right and what is wrong. Without that brain part, or certain functioning "subcomponents" of that brain section, he says, moral learning does not occur.

Knowing what we know about brain development now, he says, it is "equally wrong and incomplete" to think of deviant behavior in only social or cultural terms. Some bad behavior, particularly "systematic" rule breaking with no remorse, he says, can be primarily biologically based, such as with the children in his study. "Maybe someone had a head injury as a child or a disease; we have to think of that," he said. "We know enough now so that we should be careful in summarily dismissing something as purely biological or purely social."

At a recent annual meeting of the Society for Neuroscience, one of the most popular panels was one devoted to ethics. In what only a few years ago would have been an unlikely scene, neuroscientists debated the very essence of morality. Was a basic knowledge of right and wrong tied to the brain? Could Columbine be pinned on a faulty prefrontal cortex or a reward system gone awry? Can we blame developing synapses for that sneer on the lip or, conversely, that hour helping in the soup kitchen?

In the end, Damasio said, those kinds of questions, while now a firm part of neuroscience, are likely to be only partly answered by neuroscience.

"The debate is beginning, but it's society in a broader sense that will have to decide," he said.

And while that debate proceeds, of course, parents, in kitchens and living rooms on the front lines, often find themselves pleasantly surprised by at least a few aspects of emerging adolescence. In fact, aside from their blossoming abilities to think deep thoughts or feel for others, the most striking changes that parents mentioned to me about their budding adolescents were, in fact, about jokes.

Time and again, when I asked parents about the parts of adolescence they liked best, they would tell me that they, too, had been struck by their teenager's growing appreciation of irony and their increasing willingness to make fun of themselves. For the first time, their teenagers could place themselves in the world and look at their situation from the outside in. One mother of three boys said that she was nearly in shock—in a good way—one day when her normally sensitive son made a joke at his own expense. Coming home from school after trying out for the squash and hockey teams, he had told her with a shrug that "hockey was a long shot and squash was a long, long shot."

"It sounds like a small thing," the mother said, "but it meant that he could get outside himself; he could laugh at himself. That was a big change, a good change."

Another mother said she was surprised when her fifteen-year-old son, who still had a bit of baby fat around his middle, suddenly started talking openly about his weight, and laughing about his "lard."

When I asked the mother of the twin boys who had spent time making LSD in the high school science lab and sneaking out of the family apartment if she could name a good part of adolescence, her face lit up. "The big good thing about teenagers is their humor," she said. "When my boys became teenagers, at a certain point, we suddenly got each other's jokes, you know the dry, ironic, and sardonic jokes. To me that was the best part by far. They just started to get it. They laughed with me."

Chapter 9

SWEPT AWAY

A Surge of Hormones Swirling through the Brain

A TWELVE-YEAR-OLD GIRL: "I don't know what it is; all of a sudden when I get mad it doesn't go away anymore. It just comes and stays and I feel angry for a long time."

A thirteen-year-old boy: "Sometimes I get these confusing feelings. The other night I begged my parents to take me go-carting; I was really excited. But when I got there I didn't want to go. I just wanted to go home. I was sad and angry at the same time. I don't get where all that comes from."

Where does such teenage moodiness come from? The best guess has always been hormones, the raging hormones of adolescence. As parents, we dread them. Sneaking in during the dead of night, they kidnap our children and turn them against us just, it sometimes seems, when they can reliably tie their own shoes.

It's not fair. But is it true?

One way or another, it is. Indeed, even those who know hormones are not the whole answer do not make the mistake of underestimating them. Neuroscientist Jay Giedd, who insists that "hormones have gotten a bad rap," in the next breath will tell you that hormones may, in fact, lurk behind certain teenage behavior after all. Some of the changes that Giedd himself found in the adolescent brain, in particular the thickening of the frontal lobes, were detected in girls at age eleven and in boys at age twelve—as he says, "a direct hit on puberty." That leaves open the real possibility that hormones may be intricately involved not only in the sexy shenanigans we blame them for, but also in sculpting the basic architecture of a teenage brain.

"What are hormones doing at adolescence?" says Liz Bates, a neuroscientist at the University of California at San Diego. "They're doing a whole lot. These are violent chemicals."

But pinning down just how this plays out in brain development and teenage behavior is slippery business. There's what we all know: the pimples, the penises, the pubic hair. But can we find the moment—is there a precise moment—when hormones hijack the teenage soul?

ANYONE who's lived with a teenager, had a bad-hair-menstrual-day or a testosterone-spraying corporate brawl thinks they know the mood-altering power of hormones. Not long ago I asked an old friend, who'd just started hormone replacement therapy after a hysterectomy, how she felt to be no longer estrogen-deprived. "How do I feel?" she said. "Well, for starters, I no longer start every day wanting to commit homicide."

Andrew Sullivan, a writer and former magazine editor who is HIV-positive and injects himself with testosterone to combat fatigue, recently described how the hormone, which has such a rich lore that scientists simply call it "T," affected him. Not only did "T," his "syringe of manhood," increase his bulk—within months he gained twenty pounds; his collar size went from fifteen to seventeen and a half, his chest from forty to forty-four—but his moods became more aggressive and volatile—more macho. He found himself, "chest puffed up," in fights with strangers, irritable, testy. "Within hours, and at most a day, I feel a deep surge of energy. It is less edgy than a double espresso, but just as powerful. My attention span shortens. . . . My wit is quicker, my mind faster, but my judgment is more impulsive," he wrote in the *New York Times Magazine*.

But can these sex hormones be detected as they hip hop through a human teenage brain?

Not long ago, Liz Susman at Penn State University, a highly respected hormone researcher, set a decent trap. She gathered fifty teenagers, all of whom were sexually underdeveloped so that giving them hormones was necessary. Susman and her colleagues gave them hormone injections, estrogen for girls and testosterone for boys, for three months, then a placebo for three months. It was a randomized double-blind study, considered the gold standard for scientific inquiry, in which neither researchers nor subjects know who's getting what.

During the experiment, researchers regularly talked to parents and teenagers about their lives: how many stomped feet, how many insolent glares? And, after meticulously culling the data, Susman caught hormones kicking in the doors. "We found more aggression in both boys and girls [during months they got hormones]," Susman told me. "And this was at doses designed to approximate what normally happens in the body."

This will not come as startling news to parents, of course. In many ways, aggression is adolescence; adolescence is aggression. Banged doors, thrown books, hurled threats. Nothing new there. When I asked a friend Bill Walsh, the father of three girls, how he knew one daughter had become a teenager, he looked at me like I was mad. "How did I know when one daughter became a teenager?" he said. "That's easy. When she looked me in the face and told me to go screw myself."

Another friend, the mother of a fourteen-year-old, took me to her house not long ago to see the hole her son had kicked in the wall, angry over . . . well, it doesn't really matter. "Hormone hell," she said shaking her head in dismay. Her son was in his room, door closed, not talking.

Still, hormones aren't a solo act. In an elaborate feedback loop, hormones make behavior, but behavior also makes hormones. With teenagers, as with anyone living a life more complicated than a fruit fly in a lab jar, there's all the squishy social stuff, the very fabric of teenagers' lives that can have its own detectable impact on hormone levels.

"A lot is happening in teenagers' lives, and some things they don't like. They get pimples; they get fat; social relationships are shifting around; they aren't as tall as their friends and they don't like that. All that could be indirectly related to hormones but it's not directly hormones," Susman said.

Estrogens and Androgens

Just taken by themselves, hormones are confusing enough. Shooting through the bloodstream, hormones tweak other cells and each other. What we're talking about here are the so-called sex hormones or sex steroids, primarily the estrogens and the androgens. Estrogens come in different flavors, but one, estradiol, gets the most attention. Testosterone is the main androgen, but there are others. We think of estrogen as a female hormone and testosterone as a male hormone but, in fact, males and females produce both. The difference is in their levels: Males have roughly ten times as much testosterone as females and females make about ten times as much estrogen. Females make most of their estrogen in their ovaries, and males make most of their testosterone in their testes.

In males, testosterone fluctuates during the day, sometimes by as much as 150 percent, with lowest levels recorded at noon. It has also been shown, at least in males, to rise abruptly when they're faced with a challenge, in the jungle or corporate office, and crash when a game is lost and it's time to regroup or surrender.

Estradiol, on a more lengthy cycle in females, ebbs and flows on a monthly schedule. But this is no wimpy tide. By some estimates, estrogen increases anywhere from 650 to 4,900 percent during a month, reaching its peak around ovulation. Testosterone also is converted to estradiol in the male brain, a feat accomplished with the help of the enzyme aromatase.

In their studies, Susman and her colleagues uncovered a variety of hormonal influences beyond aggression. During the months when the girls received hormone injections, they were more withdrawn. They wanted to be by themselves, doors closed, left, if you don't mind, alone. But the hormones, just to throw us off their scent, also promoted a kind of togetherness. During the months they received hormones, boys reported more sexual thoughts, more nocturnal emissions, more episodes of "touching girls," and even a bit more sexual intercourse. Girls, too, when awash in estrogen, had more "necking" episodes and sexual fantasies, although they restrained from being carried all the way by those thoughts much more often than boys.

But tracing these changes and behaviors back to the microarchitecture of the growing teenage brain is not easy.

To see just how tricky all this is you only have to go see Art Arnold and his bird. In the back corner of Arnold's UCLA lab is a tall wire cage with a small bird hopping around inside. Arnold, a longtime hormone researcher, explains that the tiny zebra finch, with a round orange patch on its head, has the distinct plumage of a male. But over the last few years, the little macho finch has regularly laid eggs. "I think it was eight the last time I checked," said Arnold, shrugging his shoulders like a man who's been honorably perplexed by sex hormones for a long time.

In the end Arnold told me, if there's indisputable news about sex hormones in the human brain lately, it's that, while they may be hard to catch at work, the role they play is much bigger than anyone had ever thought. For years, science believed that estrogen and testosterone acted mostly on parts of the brain linked to sexual behavior, primarily the hypothalamus, the peanut-sized knot of cells in the middle front of the brain that, through its various subdivisions, regulates a mixed bag of critical functions including sex drives, ovulation, thirst, and hunger.

Using today's more sophisticated tools, however, researchers have now found receptors for estrogens and androgens sprinkled all over the human brain, in the cortex and the cerebellum, two areas associated with movement and cognition; in the amygdala, linked to strong, gut emotions; and in the hippocampus, an area important to memory.

"They seem to be all over the place," said Arnold.

So what exactly are they doing in there?

Arnold shrugged again. It's complicated.

PUBERTY IN THE BRAIN

Puberty, for starters, begins in the brain, though even today no one knows what triggers it. One guess is that levels of fat, detected and passed along by the hormone leptin, which is produced in fat cells, might be connected. Puberty takes a lot of energy and the body needs a certain level of fat reserves before it launches into this next big developmental stage. This may be why anorexic girls, or even extremely athletic ones such as gymnasts or ballerinas, sometimes do not menstruate on time. The female body may need a certain fat level and weight before puberty is triggered.

Other scientists, however, are convinced that fat and leptin levels, while a necessary gateway, are not the main initiator. Puberty, they suggest, may be set off by the natural cell pruning process that starts in certain areas of the preadolescent brain. At some point, certain genes kick in and the neurons that release the inhibitory neurotransmitter GABA in the hypothalamus, are pruned back. And with less inhibition, this theory goes, the hypothalamus happily goes back to what it was doing in the womb and infancy—revving up sex hormones.

In any case, something prompts the hypothalamus to jump-start the nearby pituitary gland. The pituitary, in turn, produces hormones that awaken the testes and the ovaries from their childhood hibernation so they can once again crank out testosterone and estrogen. The process begins in girls around eight years old and in boys about ten. Hormone levels rise steadily, culminating in the onset of menstrual cycles in girls at an average age of thirteen, and production of sperm in boys about age fourteen.

The impact of this hormonal tide, outwardly, is hard to miss. With a little help from growth hormones, also firing up at this age, girls grow an average of ten inches and boys eleven inches during adolescence. It's estrogen that prompts the cartilage cells of the hips to stretch, allowing for childbirth. It's testosterone that pushes a boy's shoulders out.

But long before that moment of puberty, hormones have been busy in a developing embryo's brain. What determines sex to begin with, of course, are the chromosomes. Chromosomes carry the genes, which make amino acids that, strung together, make the proteins that construct the bodies, boy or girl. Of the forty-six chromosomes we have, arranged in twenty-three pairs, it is the last pair, the X and Y chromosomes, that determines one's sex. Males have an X and a Y chromosome and females have two Xes. That means when a sperm fertilizes an egg, it contributes either an X or a Y, and that's why it's the father who determines the sex of the child.

Before six weeks, an embryo looks as if it could go either way, male or female. But after six weeks, a split occurs. If the embryo is a genetic male, its Y chromosome has genes that sculpt part of the lumpy little embryo into testes, which in turn start pumping out testosterone. Conversely, in embryos without the Y chromosome, there's no surge of testosterone and it proceeds to develop into a female with a uterus and fallopian tubes.

As the embryo grows, estrogen and testosterone help to sculpt different sorts of brains, male and female. Some early differences, called the organizational effects of hormones, are evident right away. Others prime the organism to respond differently when testosterone or estrogen, after disappearing for a stretch of time in early childhood, reappear at puberty. In either case, hormones are far from timid as they work in the brain; their impact is both pervasive and far-reaching. It's now known, for instance, that sex hormones can make brain cells and branches grow or disappear, make neurotransmitters excited or calm, and, working on the inside of the cell, turn genes in the nucleus on and off. Quite simply, as Cynthia Bethea at the Oregon Regional Primate Research Center, put it: "They can change the function of the neuron."

IN ANIMALS, the swirl of hormones has been successfully tied to several specific structural differences in male and female brains. And those structural differences have been traced to distinctly different behavior. Art Arnold, along with Fernando Nottebohm, did a landmark study in 1976 in which they found that in birds where only males sing—zebra finches and canaries, for instance—the vocal area in the brains of male birds, awash in testosterone, is six times larger than it is in female birds. And if you give a shot of testosterone to a young female zebra finch, she gets a bigger song spot in her brain and she starts to sing, too.

In rats, too, there are clear lines between hormones, brains, and behavior. If you deprive male rats of their testosterone at a certain early sensitive period, they often rudely ignore the female rat in the next cage, no matter how attractive she might be. And if you give testosterone to a female rat, she not only fails to assume the submissive position of lordosis, in which she sticks her fanny in the air to accommodate an approaching male, but sometimes grows her own penis.

Jill Becker, a psychologist at the University of Michigan, has found ways that estrogen, too, works in certain parts of the brain, not only inside the cell but on the outside as well, a quicker process.

In fact, Becker thinks the reason some earlier studies failed to find a direct impact from hormones is that they looked too late. She believes hormones can act in a more immediate, minute-by-minute, rapid-fire fashion on the outside of the cell. And if so, she says, their influence all over the brain could be "mind-boggling."

In her careful rat studies, Becker has shown that estrogen interacts with dopamine in females in just this way. Dopamine is an active and powerful neurotransmitter. Most of the drugs of addictions, as mentioned earlier, including cocaine and amphetamines, increase dopamine and, in that way, make the world seem better, brighter. Rats will cross a lot of hurdles to get a fix of dopamine. On the other side of the coin, Parkinson's patients, with depleted dopamine, sit stiff and immobile, unable to muster the ability to move.

In her experiments, Becker has shown that estrogen increases dopamine in the basal ganglia, the inner brain structures that help initiate movements. Estrogen, she says, blocks GABA, the main inhibitory chemical in the brain. And when inhibition is reduced in areas where neurons release dopamine, you get more dopamine. Even in a test tube, Becker has found that if you take parts of the basal ganglia and add a dollop of estrogen, dopamine increases.

Becker has shown this in the real world, too, at least in the rat real world. If a rat is running high in estrogen, at the peak of her estrous cycle, for instance, she will do much better on a balance beam, a movement-related task. And, perhaps more important, she will also be better at timing her episodes of mating. In the wild, female rats run away for a time after males mount them, allowing for a complex system of hormones and neurochemicals to work effectively and increase their chances of becoming pregnant. Studies have shown that if a female rat times it right—and she will jump over big obstacles both to get to a male rat and to get away from him and give her body a rest—she can increase her chances of reproducing by as much as 90 percent. And Becker has found that the female rats time it best when estrogen and dopamine are high, when they are most alert to what is going on. "The dopamine tells them to pay attention, this is important. Get it right," says Becker.

And the process may very likely be at work in humans as well. If estrogen is revving up dopamine in adolescents, and that dopamine makes the world seem brighter overall, teenagers, Becker says, may be dealing with a "more vibrant world," and that can have real implications for their behavior. Just at a time "when they are trying to figure out their place in the world," she says, teenagers may be walloped with an outsized, almost psychedelic view of the world around them—reds are redder; blues are bluer. Their world may be more aglow, more exuberant. But such heightened experiences can work the other way, too. If a teenager is feeling sad or depressed, he or she may experience it as more awful, more distressing. A mother's frown is deeper, a sideways glance at a school dance taken the wrong way is the end of the world. Dopamine may be painting walls of the mind bright purple, turning up the inner radio and goading: "Go grab it, Do something! Jump!"

MALE AND FEMALE BRAINS

Perhaps not surprisingly, however, the study of how exactly hormones play out in humans has been a bumpy and contentious road.

As Deborah Blum lays out in her book *Sex on the Brain*, research into any hormonally pushed sex differences in the human brain was stymied for years, not so much by science as by politics. It's an issue that makes people angry and defensive. It was none other than famous nineteenth-century neuroscientist Paul Broca, the man who first found a distinct language area in the brain, who declared, Blum says, that women were slightly dumber than men because their brains were made, from day one, in a smaller size.

Females, on average, have brains that are a full 15 percent smaller than male brains, about three ounces smaller. As Blum quotes Broca on the subject: "We might ask if the small size of the female brain depends exclusively on the small size of her body . . . But we must not forget that women are, on the average, a little less intelligent than men."

Granted, this was one hundred years ago, but things weren't much better by the 1960s when, as Blum writes, "the pendulum of scientific thought had swung to the opposite extreme from Paul Broca and, except for the annoyingly persistent difference in overall size, the popular belief became that men's and women's brains mirrored each other."

Now, for the most part, we're past all that. The thought today is that males and females do, on occasion, act differently—not better or worse, not stupider or smarter, but differently, although some of those differences are continually called into question. The hunt is on now to find where in the brain those differences come from—is it from hormones, from brain structure, from experience?

Generally, IQ scores are relatively even between the sexes. But there are persistent reports that males and females handle certain tasks differently, with many of the discrepancies emerging at a time when hormones crash onto the stage again—at puberty.

Females around the time of puberty start outperforming males on certain verbal tests, such as name as many words as you can think of that begin with the letter *d*. And males, beginning in adolescence, outperform females on certain spatial tests, such as here is a row of blocks arranged in a twisted pattern, now rotate those blocks halfway to the right, what does it look like?

These differences in performance between the two sexes are statistical; there are plenty of girls who rotate better than boys and plenty of boys who think of *d* words faster than girls. And there are other problems with a strict adherence to this line of thought. While male rats may make fewer mistakes in negotiating mazes—a spatial task—Becker believes that may simply be because females, with higher levels of that motivator dopamine, are more active, trying more things and making more errors along the way. "They take more blind alleys," but they get there just as fast in the end, she says. And as Blum points out, one study found that males and females might simply do mazes, not better or worse in the end, just differently. In one series of experiments, female rats got hopelessly lost if researchers took away bright-colored landmarks in mazes, but the change made no difference to male rats. Studying human college students trying to navigate a computer maze, researchers also found such differences. Changing the geometry, making one arm longer than the other, "threw the men," Blum writes. But it was removal of the landmarks that flummoxed the girls.

This suggests a difference in strategy, not performance. A female giving directions to her house might say: "Turn right at the big white church." A male, on the other hand, might say: "Go 1.45 miles and turn south-southwest." (In fact, I think I know him.)

SO, EXACTLY where do those discrepancies come from? Are we still calling on boys more than girls in math class? Are boy and girl brains, sculpted by different molecules, irrevocably different—vastly different—from the get-go?

Despite their squabbles, scientists have agreed on a few areas in the brain that do seem to differ by gender, the so-called sexual dimorphisms. In several areas of the hypothalamus—some related to sexual behaviors, some related to who knows what—male brains are clearly bigger. On the other hand, certain segments of the fiber bundles that connect the two cerebral hemispheres are bigger in female brains. Is this why males are generally more sexually aggressive than females, or why women have been labeled "whole brain" thinkers? One brain-imaging study found women rhyme with both sides of their brains, while men use only one side. There's also evidence that after suffering certain strokes, women do better than men. Could that be because their connecting fiber bundles are bigger and they can call on both sides of their brains for repair?

AMYGDALA VS. HIPPOCAMPUS

Several years ago, Jay Giedd spotted a few sexual dimorphisms in adolescence, the first detected in the developing teenage brain. Giedd found that the amygdala, the tiny wad of cells in the center of the brain that helps prompt those gut, meet-me-outside impulses and which is also filled with testosterone receptors, grows faster in teenage boys than in teenage girls—perhaps one reason a higher proportion of boys bloody-nose their way through sixth grade

Giedd also found that the hippocampus, where certain memories are formed and which is littered with receptors for estrogen, grows faster in adolescent girls than boys—perhaps explaining, in part, why sixth-grade girls are so much better at memorizing spelling words.

Most recently, Giedd discovered that the lowly cerebellum, that lump near the top of the neck that science has never given much thought to, is the most different, the most sexually dimorphic part of the brain, as much as 14 percent larger in adolescent boys than girls. By studying twins, he also found that the cerebellum is the least heritable section. That means this little brain pod, linked to movement and certain types of social cognition, is being molded, in males and females, by forces on the outside as well as the inside.

"I'm guessing here," Giedd said, "but I'd say that in evolutionary terms, there was more pressure on men to develop parts of the brain that are connected with spatial skills like hunting and throwing a spear, all the kinds of skills that are thought to involve the cerebellum."

Giedd, along with many neuroscientists, believes that, while the connection between form and function can be elusive, one way or another, in the brain size does matter. (Not to worry, there is some evidence that although female brains are smaller, they may run "hotter," or be more efficient.)

Giedd uses the analogy of two schools to make his point. If you have one high school with spaces for fifty students and applications from sixty, most get in. But if you have a more competitive high school, one with spaces for fifty students and applications from a thousand, that school will get a better group of students because it can be more selective.

Giedd believes the brain may act the same way at adolescence. It's at adolescence, after all, when full-scale pruning of brain branches occurs—more than 50 percent of the neural connections are eliminated in certain areas. If, because of hormones, a certain brain part is bigger to begin with, there might be more connections to choose from and those that are left after pruning would be the best and the brightest.

For instance, a smaller basal ganglia, the set of structures devoted to movement and a few other things, has been associated with attention deficit disorder. Boys tend to be diagnosed with ADD more than girls. There could be lots of reasons for that, but it's also true that boys, on average, have smaller basal ganglia than girls. "Perhaps when it comes to ADD," Giedd says, "girls have an extra margin for error."

ADD is not the only disorder that acts this way. Giedd points out that studying how female and male brains develop, and how they are different, is important because "nearly everything we study" in terms of neurological diseases differs by gender. Girls become depressed more frequently and earlier; rates of ADD, Tourette's syndrome, and schizophrenia are higher and tend to start earlier in boys. Why?

EXACTLY when differences arise in the brain is not completely nailed down, though many are likely to begin with the early organizational effects of hormones, specifically the early exposure or nonexposure to testosterone.

One good example of this in humans is the so-called CAH girls who, because of a genetic defect, end up with a large rinse cycle of androgens in their brains before birth. Girls with this defect, called congenital adrenal hyperplasia, lack an enzyme that helps make sufficient cortisol to shut down androgens produced in their adrenal glands in the womb. Consequently, the girls get extra male androgens prenatally. If the case is severe, the girls are born with a clitoris that's more like a tiny penis, requiring surgery and medications to correct. Milder forms are evident mostly in behavior.

Sheri Berenbaum at Penn State University, who has studied these girls, says the extra androgen elicits nontraditional female behavior you can actually watch in CAH girls as young as three years old. In a series of careful studies, Berenbaum found that the CAH girls, left in a playroom alone, consistently picked fire and dump trucks and other traditionally masculine toys to play with, rather than dolls or furniture. As they got older, moving into adolescence, Berenbaum found that these androgen-rich girls, while they think of themselves fully as girls and had been raised in a girl-culture, consistently performed better on spatial tests, shied away from makeup and baby-sitting, and later tended to pick professions more normally associated with boys: engineering, flying, and architecture.

In these cases of CAH, it seems clear that early exposure to hormones in the brain has significant impact on how the brain is organized for life and how a human with that brain acts in adolescence and beyond.

"You would think that as [the CAH girls] got older, because of the impact of socialization, peer pressure, and the hormones of puberty, that the impact would be smaller," Berenbaum told me. "But the differences stay. We find these girls continue to fall somewhere between the typical girls and the typical boys. That suggests to me that the amount of androgen you are exposed to prenatally has an effect on some part of the brain that gets you interested in boy toys and boy activities."

So what you are interested in depends, in part, on your androgen level as a baby? I must admit I still hate hearing this, but there it is and Berenbaum, an avowed feminist, isn't apologizing.

"These differences don't mean that someone is inferior or superior," she said, detecting my dismay. "It doesn't explain why women are paid less. That's a social thing. And just because girls and boys are different doesn't mean a girl can't be president. It just helps us understand better how the brain works."

In any case, there's also evidence that sex hormones can make brains different at other times as well. Jill Becker, for instance, found estrogen affects dopamine levels during every estrous cycle throughout a woman's life. Just a few years ago, researchers found that hormones alter brain structure in adults, too. During the estrous cycle of the female rat, when estrogen was high, the brain branches—the dendrites in the rat's hippocampus—grow and shrink, all in the space of four days. No one knows why exactly. Some have speculated that since female rats must mate in a small window of time, they have to be able to range as far as possible from their nests and still make their way back again, a memory task that the hippocampus may well help with.

And then there's the latest wrinkle of all: new baby neurons. The neuroscience world was dropped on its dogmatic head just a few years ago by reports that the adult brain, which in this case includes adolescents, continues to produce a regular stream of new neurons, at least in the hippocampus and possibly in other areas. In many ways, the grown-up brain has to be immutable, or we would never remember where our socks are. But in the last few years, neuroscientists have been finding that the brain continues to change—it is plastic—throughout life, with some changes clearly tied to hormones.

"Every year, when I go to the annual neuroscience meeting, I have to laugh," says Marc Breedlove, a neuroscientist at Michigan State University. "Every year the brain gets more plastic than the year before."

And with new clues has come new interest. After decades of neglect, the study of how hormones affect the brain at every age, and in particular at adolescence, is now what Judy Cameron at the University of Pittsburgh describes as a "hot, hot area." One of the biggest discoveries came in 1996 with the detection of a second type of brain receptor where estrogen can dock.

For years, hormone hunters had looked for only one type of estrogen receptor in the brain, ERalpha. And they didn't find that many outside the expected area of the sex-regulating hypothalamus. There was, Cameron says, a small group of scientists who insisted that hormones had to be major players throughout the brain. But no one could figure out how, if that were true, it was happening.

"Without a clear mechanism, most just thought the idea was absurd," Cameron says.

But every now and then, the big thinkers are proved completely wrong. With the discovery of a second type of receptor, ERbeta, scientists looked again and, indeed, ERbeta receptors were all over the brain, in brain cell after brain cell, waiting for estrogen to fly on by. That means, Cameron says, that there's now an understandable "scientific mechanism" to explain how hormones might work throughout the brain to influence everything from emotions and cognition to driving too fast down dark roads. And since some testosterone in males can be turned into estrogen, this hormonal mechanism probably affects teenagers of both sexes.

"We know that there are a lot of changes in behavior in adolescence," Cameron says. "All of that could be influenced by hormones."

ESTROGEN AND SEROTONIN

Some of these new discoveries, in fact, have already changed how we think about teenagers and another powerful neurotransmitter, serotonin. Produced at the base of the brain by neurons that extend tentacles all the way to the forebrain and down to the spinal cord, serotonin affects a mudroom full of brain activities such as mood, pain, and hunger. Generally, serotonin is a calming influence—in fact, most of the new antidepressants work by allowing serotonin to wash through synapses longer. Low levels of serotonin have been associated with depression in both teenagers and adults. And for years, scientists suspected that estrogen had an effect on the serotonin system in the brain, but they couldn't figure out how the mechanism worked. They couldn't find estrogen anywhere near serotonin.

Then, after the discovery of the new estrogen receptor, ERbeta scientists like Cynthia Bethea of the Oregon Primate Center went back to their monkey brains and looked again. And there it was. Deep in the nuclei of brain cells that produce serotonin, they discovered the newly found estrogen receptor.

"Here we were waving our arms for ten years about this, but we couldn't find anything and then there it was," Bethea told me. "It was very exciting."

How the flood of estrogen might be affecting serotonin in normal human teenagers; how together, they might wreak adolescent havoc, isn't certain. As Bethea stresses, both estrogen and testosterone operate in complex feedback loops, with differing levels likely to elicit different behavior. Get the right amount and you might feel good. Get too much and you might get the jitters or, given their widespread influence, change the very structure or operation of your brain.

"Teenagers are driven by hormones and the deluge and fluctuations at puberty are very destabilizing," says Bethea.

And so how, in the end, can we sort through that destabilizing storm? Is it possible to separate the effects of the hormonal tides on the inside from the environmental winds of MTV and hovering mothers, battering on the outside?

One scientist wrestling with it is Alan Booth, a long-respected testosterone researcher at Pennsylvania State University. In a study of four hundred stable, middle-class families in central Pennsylvania, Booth and his colleagues are testing testosterone levels in teenagers, male and female, on a regular basis. And so far, they've found that the impact from the environment, the "wind," trumps testosterone, the "tide," in every case.

When parent-teenager relations are poor, high-testosterone sons are more likely to engage in risky behavior, such as skipping school, sex, lying, drinking, and stealing. Low-testosterone sons with poor parental relationships are more likely to be depressed, Booth found. (No one knows how low testosterone may bring on depression, but one guess is that it could result in low estrogen, which in turn could mean lower serotonin and depression, says Booth.)

At the same time, low-testosterone daughters who had poor relations with their mothers are more likely to do risky things, while low-testosterone daughters who had bad relations with their fathers are more likely to report signs of depression.

But the good news is that among teenagers—both girls and boys—with good relationships with their families, high and low testosterone levels don't seem to matter at all.

Attempting to sort out these varying influences, neuroscientist Marc Breedlove has turned to the rat. In a series of careful studies, he has shown how environment and behavior influence hormones and brain structure, and how hormones and structure, in turn, influence behavior—the full loop.

Breedlove studies the rat amygdala, that small structure deep in the brain that helps make split-second fight-or-flight responses—the same structure that Giedd found was growing faster in teenage boys than teenage girls. Specifically, Breedlove looked at one particular part of the amygdala, the medial amygdala, which had been shown to be involved in a rat's response to the airborne odors, called pheromones, that stimulate sexual responses and are linked, through levels of male androgens, to the intensity of rough-and-tumble play.

As Breedlove points out, one of "the most robust differences" found between males and females, from rats to humans, is the level of rough-and-tumble play. Males simply do more of it. And males running high in androgens engage in more rough play than males on the low side.

Working with graduate student Bradley Cooke, Breedlove found a couple of intriguing things about how hormones, sex pheromones, and play interact to change both rat brains and behavior and back again.

First, to their surprise, the researchers found they could manipulate the size of the adult rat amygdala through testosterone. If male rats were castrated, their medial amygdalas shrank. And if female rats were injected with testosterone, their amygdalas grew by 50 percent.

Next, Breedlove wondered if they could influence that structural change by behavior and environment alone. Male teenage rats were put in an isolated cage and allowed no rough-

and-tumble play. The rats without play developed a smaller medial amygdala—their neurons were smaller. And later on they were far less responsive to the pheromones that stimulate sexual activity.

So there you go. Behavior changed hormone levels and brain structure, and the altered structure then determined how that animal behaved.

Breedlove thinks the human brain is so interactive that you can't discount experience anywhere along the line. While early hormones make a penis, just having a penis will bring different experiences that are likely to alter hormone levels, brain structure, and behavior later on. It's not simply that structure equals behavior or hormones equal behavior but that they all tumble around in the same gym, pushing and shoving on each other.

"The brain is not made the way a human engineer might make it," said Breedlove.

"Testosterone early in life may set the stage in the brain so that at puberty certain programs kick in. But those programs, along the way, can also be changed by experience. A whole lot happens between the time when an animal is born and when it takes on its new role of being something that reproduces."

Chapter 10

The Neurons of Love

How the Brain Gives the Heart Away

Sadie Oliver was seventeen years old, half-asleep in marketing class, when it happened. Sitting in a small cold room on an autumn morning, the students had introduced themselves, the teacher had started to lecture, something about product placement. Or so Sadie thinks. Mostly, in her early-morning fog, she found herself focused on the dark-eyed boy sitting across from her.

"Brett, he said his name was Brett and I was saying it in my head over and over so I'd remember it," she recalls.

Through friends, they got together. Within days, they had sex. Now, they're getting married.

"I couldn't get him out of my head," she said.

Jesse was only thirteen when it hit him. He saw her at a field hockey game. She had a nice smile. For the first time, Jesse wanted to be with one girl all the time. After a few weeks, he wrote her a poem. She hugged him and he knew she felt the same way.

"I started singing and dancing, literally," remembers Jesse, now a sophomore at Dartmouth, who said the relationship went on to be quite intense. "I was on my bike but believe me, I was dancing."

Singing, dancing. Dancing, singing. Teenage love. If there's one ritual rooted firmly in adolescence, it's the ritual of teenage love. Luscious, lusty teenage love.

But how much, really, do we know about it? Do feelings of love and sexual attraction simply surge up on schedule, washed up with the tide of testosterone and estrogen?

Helen Fisher, an anthropologist at Rutgers University who has studied love for years, says what strikes her is how similar love is wherever it rears its lovely head. Across cultures, across literature, she says, the love-struck tread paths so alike they could be plotted together on a graph. And to her, that means that love—and not just sex—must be an involuntary set pattern of behavior driven by basic biology, in particular the biology of the brain.

Fisher's idea is that love comes in three distinct and largely universal acts: lust, attraction, and attachment. And she believes that different stages of love correspond to different parts of the brain. That is, different structural areas, different hormonal and neurotransmitter systems, kick in depending on which stage of love one is in.

"This is a system that can operate profoundly," she says, echoing a thousand years of poets. "I'm coming even more to appreciate how much it can mean. I see students lose whole semesters just because they're in love. And it happens in stages."

What Fisher sees as the first stage, lust, would roughly correspond to the point when Sadie first saw Brett across the room or Jesse saw that smile. In this stage—the craving phase, the stage where energies become focused on one person—Fisher believes testosterone is to blame, in males and females. (There's evidence, however, that estrogen also works to drive sexual attraction. Female rats, for example, must be in estrous and running high in estrogen or they can't, physically, even begin the act of sex. Even female primates, which don't have to be in heat to have sex, nevertheless become more intimate when estrogen kicks in. Some studies have shown that human females, too, if they aren't worried about getting pregnant, initiate sex or think about sex more in the middle of their ovulation cycle when estrogen in highest. And there's also the intriguing idea that if estrogen increases dopamine, it could easily elicit grab-that-one thoughts in girls.)

But Fisher is convinced of testosterone's role by evidence showing that when some middle-aged women whose sex drive has waned, add a touch of testosterone to their hormone-replacement therapy, they became more interested in sex. Some studies of young girls have found that sexual thoughts and activity progressed to actual intercourse more often when testosterone levels were higher. "You can't rule out estrogen, but I think really it's more testosterone," Fisher says.

Stage two, says Fisher, invariably involves attraction. She thinks this stage is connected to that severely overworked brain chemical dopamine. Fisher believes that it's dopamine, combined with the brain stimulant norepinephrine, that produces the love high. An increased dose of those two chemicals, combined—as the brain seeks the status quo—with a corresponding decrease in the calming neurotransmitter serotonin (whose low levels have been associated with obsessive thinking) and, voilà, the pavement doesn't stay beneath your feet.

"This is the giddy high, the butterflies in the stomach stage; the time when you can't get [the other person] out of your head," she says.

And stage three? Stage three comes later. As Sadie describes that phase, her love affair, after time, was one of "still strong feelings, but deeper, calmer, less wow."

Fisher suspects this less-wow stage is linked to oxytocin in girls and vasopressin in boys, two multiuse hormones produced by the pituitary, the pea-sized gland at the base of the brain. Both have been linked with bonding behavior. (Among its many jobs, oxytocin brings on uterine contractions at labor and helps eject milk after the baby comes. Vasopressin helps the body retain water, particularly in emergencies. Although both are difficult to track in humans, if they're injected into brains, rats cozy up to the next rat that walks by and female monkeys get more google-eyed over monkey babies. And oxytocin, linked to rhythmic behavior such as stroking an infant, is also released at orgasm.)

Of course, having such theories is one thing, proving them is quite another. But Fisher is determined. Working with Arthur Aron, a psychologist and longtime love researcher at the State University of New York at Stonybrook, Fisher has embarked on an unusual project: She is looking inside the actual brains of adolescents in love.

Day after day, Fisher and Aron strap the young and smitten into brain-scanning machines and, as they gaze at pictures of their boyfriends and girlfriends, monitor their brain activity.

They cannot, with current technology, see the chemicals coursing through the brains of their college student subjects. But they're trying to identify the brain areas known to respond to those chemicals and see if they click on.

Although the research is not complete, the researchers expect that the love-struck brains will ignite in similar areas. In particular, they are looking to see if the brain's reward circuit is activated—the part that's stimulated when we win at poker, eat a piece of cake, or sniff cocaine and is turned on by dopamine.

One of the big unknowns about love, says Aron, is whether it's an emotion or a motivation. This may seem like a highly technical and theoretical question, but in fact it's quite fundamental. When asked to name an emotion, most often people list love as one of the first. But love is different from other emotions. Anger and sadness, for instance, are associated with facial expressions that you can recognize. Love has no particular facial expression linked to it. What's more, unlike other emotions, Aron says, love involves a range of emotions, including anger and sadness and even guilt. And if the research shows that those in love activate the reward centers of the brain that are stimulated by dopamine, it will be a good clue, Aron says, that love is a motivator, not an emotion.

In fact, dopamine has already been shown to be involved with attraction and reward in humans. Over the past several years, Bruce Arnow and his colleagues at Stanford have persuaded a group of young male college students to lie in a functional magnetic resonance machine and watch porn films. As they watch the movies, which depict a variety of sexual activities, a specially designed device measures their level of what Arnow calls their "peripheral response"—that is, how erect the penis gets. The simultaneous brain activity is recorded and compared with how the brain acts when the young men watch more neutral scenes, such as waves on a beach or baseball.

And they find that, as the men respond sexually, their brains light up in several intriguing areas, including the caudate and the putamen, inner brain areas rich in dopamine receptors and big players in the brain's reward circuit, where the urge to get more of that wonderful feeling runs rampant.

"We haven't really known how the brain is involved in sexual arousal in humans," said Arnow. "But this study, I think, shows that it has to do with dopamine and it has to do with reward." For one reason or another, he says, perhaps confirming what many already thought they knew, "the brain sees sex as a reward."

Frisky, Risky Love

What might all that mean for the average teenager? Teenagers are attracted to risk. And love might be just the kind of "risk" their brains like. Dopamine, which responds to risk and novelty, would excite them, get them off the couch, or on, as the case may be.

Although teenagers are hardly the only ones who fall in love, it's an activity that is certainly aroused in the adolescent years. To Aron this should not be a surprise. His research has repeatedly shown that people fall in love more readily if they're already in a physically aroused state. That doesn't mean only a sexually aroused state, but any activity that gets the blood running high. For instance, two people who meet in a scary place, like on a high suspension bridge, or who think they're going to get an electric shock in a lab experiment, or who've been running on a treadmill or even watching a comedy tape, are much more likely to become attracted to each other.

Given that, Aron says, it's hardly surprising that teenagers fall in love fairly often, since they're often in "highly arousing situations." Adolescence, he says, "makes everything large." Teenagers are often highly stimulated, on an emotional roller coaster.

There's been a rash of studies lately raising concerns over young teen romance; one study found that teens, particularly girls, who got romantically involved at age twelve or thirteen were more likely to be depressed. Overall, Helen Fisher does not believe that teenagers are generally any worse at love than the rest of us. "Let's face it," she says, "no one does a good job at this." Still, because she believes love is rooted in brain systems that are still developing, teenagers may be more impulsive—and get into more trouble—in this area, as in others.

"The prefrontal cortex develops slowly. That could mean a lot," Fisher says. "They have strong drives but not the brain power or the experience to go with them."

Some scientists, in fact, now believe that sexual urges and brain development must eventually get into sync for things to work well at all. While the two systems obviously influence one another, they may also evolve along their own tracks, with effective reproductive behavior emerging only when they are both fully developed.

Kim Wallen at Emory University, who runs a long-term study of more than three hundred rhesus monkeys—which are a bit easier to watch in a systematic way than a troop of ninth graders—has found hints of this pattern in monkey adolescence. The female monkeys that, for whatever reason, go through puberty a year earlier than others have regular monthly cycles and seem reproductively mature. But initially they don't care about sex and they don't get pregnant. Wallen says that this might be because the young females don't, at first, know what they're doing. While it's up to female rhesus monkeys to initiate sex, he says that the young ones can "get excited and overdo" at first. And the males, already fearful of wary-eyed matriarchs who dominate the rhesus monkey colonies, get the jitters and steer clear.

All female teenage monkeys tend to be awkward and shy. Once, when Wallen was watching one young female sidle up to one of the big male monkeys, he had to laugh. As she reached her small hand over to touch him, it was shaking wildly.

"You can imagine what it's like," Wallen says. "These young females have been living with these male monkeys their whole lives, and then one day they wake up with an urge to go over and sit near them and groom them. They must get there and wonder 'Why do I want to do that?'"

Still, Wallen thinks socialization isn't the whole story. Early maturing monkeys may not initiate sex because their brains are not fully ready. While they appear reproductively mature, they are not cognitively mature; certain brain connections aren't wired up yet. So while they have sexual urges, they don't have the foggiest clue about what to do with them because the brain structures that are set up to interpret those signals are not finished.

"It might be that the control of their neuroendocrine system is mature but the brain area that responds to that system and initiates sexual behavior is not," says Wallen.

This has implications for human early maturers as well, a group known to get into more trouble generally than those on a more typical timetable. According to Wallen, their reproductive system could be "turned on before other brain systems to control it are in place," a recipe for disaster. In the end, Wallen believes we may find that hormones and sexual behavior develop more independently than we think, with a lot riding on the development or nondevelopment of the teenage brain. "My suspicion is that we'll find some of these things are linked to hormones," Wallen says. "But the more exciting stuff may be that there are major maturational changes in the brain that start in fetal development and then go on their own timetable."

Some researchers who work with teenagers think we've seen this already. In the long-term look at underdeveloped adolescents he did with Elizabeth Susman, Jordan Finkelstein, a doctor and behavior professor at Penn State University, said he was struck not by how much sexual behavior was tied to hormones, but how little. (In the study, sexually immature teenagers were given estrogen and testosterone for three months at a time, and then taken off the hormones, to determine how much of their behavior changed with rising levels of sex hormones.) These were teenagers, Finkelstein says, whose friends were all having sex; there was little reason, once they had functioning parts, for them to refrain. But he said, while some sexual-like behavior increased—"they were thinking about sex"—there was very little actual sexual intercourse among the group, even during the months when they were given hormones. Why?

Some may have avoided sexual relations for a variety of reasons. But because their peers were all sexually active, Finkelstein thinks their reticence stems from an immature brain. Even a higher level of hormones was not enough to push them toward increased sexual activity. Perhaps the neurological systems that control such things didn't have the requisite experience. Their brains were not yet up to the task.

"People tend to think that one day kids are doing nothing and the next day they are having sexual intercourse. Maybe that happens for a few. But for most, it's a gradual process," says Finkelstein. "First, there's solitary sexual behavior, then you start to extend yourself to others, try to touch others, put an arm around a waist, brush against a breast or a penis while you're dancing. That takes place over several years and maybe hormones are involved and maybe not. Maybe these kids didn't do anything because they hadn't gone through the behavioral stages, the stages that are primarily learned in your brain."

Other Hormones

Bob Rosenfield, an endocrinologist at the University of Chicago, says that when considering sex in humans, you can never underestimate the power of the brain. If the brain is underdeveloped, it might keep things from happening; once developed, it can keep things going. Although testosterone and estrogen are involved in sexual behavior, he says it's amazing how much can get done without them. Even men with extremely low testosterone levels often continue right along with active sex lives.

"We humans have a marked capacity to learn sexual behavior; we almost get addicted to it and we can keep at it," Rosenfield says. As he points out, even four-year-old boys, despite their seeming lack of testosterone, sometimes get erections in their sleep—fairly good evidence that, as far as some sexual actions are concerned, there may be a solely "neural route."

Martha McClintock, a longtime hormone researcher at the University of Chicago, suspects all the routes, neural and otherwise, start their journey earlier than we think. McClintock, in fact, thinks the processes of love might start long before pimples. Analyzing studies of older adolescents who thought back to when they had their first crush, McClintock found that across the board—male and female, heterosexual and homosexual—all remember having their first true crush at about age ten, in the fourth grade.

As McClintock sees it, it could hardly be related to the regular well-known pubertal increase in hormones from the gonads. If it were, girls, who reach puberty two years before boys, would have their first big crushes much earlier than boys. It could be related to brain growth in terms of learned behavior. That would mean that the parts of the brain that tell us to mimic grown-ups somehow wake up and push us toward aping parental behavior. But that can't be right, McClintock reasons, because this age-ten phenomenon shows up not only in heterosexual kids but in gay kids, who had fully heterosexual parents.

So what is it? McClintock thinks it's related to hormones other than testosterone or estrogen, ones that are less well known but appear in boys and girls beginning as early as age six or seven. These are androgens, and they arise not from the testes and ovaries but from adrenal glands. At this age, the adrenal glands become mature and begin pumping out a form of androgen known as DHEA, which recently has become popular as a dietary supplement that's supposed to increase energy and whose metabolism leads to testosterone and estrogen.

The name for this early rush of androgens, whose levels rise significantly around age ten, is adrenarche, and McClintock and a few others now think it has been severely underrated as a player in puberty. "The pediatric endocrinologists know about it, but it's not really out there in the popular consciousness; we don't think of kids as hormonal at that age," McClintock says.

Some preliminary studies also point to adrenarche as a possible starting point for a variety of teenage ills from aggression and conduct disorder to bipolar depression. Could it also be part of the hormonal engine of puppy love?

McClintock says it's important, however, not to leave the brain out. Researchers, she notes, found that age ten marked the first *remembered* crush, the first one that made a big impression. Most people have a vague sense of some tentative boyfriend or girlfriend in kindergarten, or even preschool. But McClintock believes that the age-ten crush may show up more clearly on the love radar because of brain connections, newly developing at that stage, that allowed stronger memories. And those connections could easily be brought about by hormones.

"Hormones can push things, connect areas of the brain that have to do with paying attention. And so when that happens, you remember," she says.

In any case, McClintock has no doubt that hormones of one sort or another are silently pushing teenage behavior even if we are not fully aware of it. In part, her strong beliefs come from the powerful effects of other unseen chemical players. It was McClintock who, when she was just twenty-three, published the impressive study, based on what she'd noticed living in the dorm at Wellesley College, that suggested that women who lived together, reacting to hidden chemicals, begin to menstruate in sync, something that the male scientists at that time knew happened in rats in their form but had never thought happened with human females, or had never bothered to ask.

Love Is in the Air

More recently, McClintock has gone even further. In a study where she wiped a tissue with the sweat from one set of women under the noses of another set of women, she found she could reset the menstrual cycle of the second set of women. In woman after woman, McClintock was able to change the time they ovulated.

It was a startling revelation and McClintock is convinced that the mechanism behind it is the barely detectable odors that send signals between animals, the pheromones, the same chemicals that, emanating from a female rat, tell a nearby male rat she is ready and willing to mate.

Pheromones make up an airborne signaling system that some, including McClintock, think is hard at work in humans as well. Studies of other animals have shown that pheromones, sniffed from urine, help animals recognize which potential mate has an immune system that is least like its own, perhaps giving any future offspring more varied and stronger protection. Comparable studies have found that college women, given a choice of sweaty T-shirts, will rate as "sexiest" the T-shirt from the man whose immunity genes are the most unlike her own. A similar T-shirt study conducted by McClintock and her colleagues last year found that women picked men as sexiest whose genes were similar to their father's, but not *too* similar, a reaction that could have evolved so that women would avoid mating with close relatives, but retain certain genetic combinations that already had proved valuable in their particular environment.

Using pheromones in mating, in any case, is doubtlessly important. As Deborah Blum points out in her book *Sex on the Brain*, one researcher found that even the lowly broccoli plant has fifty different kinds of genes to keep it from mating with an overly similar broccoli.

McClintock said she is unsure why pheromones might signal human women to change ovulation patterns. At some point in our history, and in places now where resources are scarce, she said, it might be important for females to recognize another's ovulation status. Perhaps it has to do with an abundance of food, making it a good time to have a baby, or perhaps there's been a drought, and it's not.

McClintock is not saying that the average teenager acts like broccoli (although the comparison is tempting), but she does believe in the backstage workings of any number of unseen chemicals. Although we don't like to talk about it, she says, even young children are sexual beings, pushed and pulled by fluctuations in hormones most of us hardly know exist—and long before a deeper voice asks a breast-budded beauty to go to a movie.

"Sexuality develops in stages just like walking," McClintock says. "You don't just go from lying around to dancing and running. It's not like Mel Brooks's Two-Thousand-Year-Old Man, who just wakes up one day and says 'Hey, Morty, I think there's some ladies here.' It happens in steps. We just don't like to think about it."

On the contrary, in our culture anyhow, most milestones of emerging sexuality are kept under the covers, certainly not cause for big parties with ponies and balloons. But maybe if we took note of these subtle markers along the way, we might not be so taken aback by puberty when it springs forth full-tilt. Perhaps we would not go through what often seems an abrupt transition, for parents anyhow, from milk and Oreos to oral sex.

"We don't have a ceremony for the first pubic hair," said McClintock. "Well, why not?"

Chapter 11

Wake Up! It's Noon

How Biology Shuts Off the Alarm

Not long ago I was talking with Cynthia, an old friend who was visiting with her daughter, Joanna. We spoke about middle school. Joanna had just survived middle school and had moved on to high school. At that point, my two teenagers were still in middle school and so we talked about who had come up with such a silly idea. Who was the peanut-brain, we wondered, who thought it was a good idea to lump hundreds of pouty, pimpled thirteen-year-olds together in a small sweaty space?

"Maybe they should just get rid of it and have a middle school dedicated to national service or something," Cynthia suggested. "Get them out there doing something."

I agreed. "How about a middle school for whitewater rafting or mountain climbing or . . . "

Joanna, listening to us, could no longer keep still.

"Don't you guys get it?" she asked, rolling her eyes in that special teenage way. "What teenagers need is a middle school for sleep. The National Middle School of Sleep. That's all we need. We all need more sleep."

As I write this, it's noon on Saturday and my own two teenagers are still asleep. One is curled up in a small fetal ball, under a comforter, no sign of life. The other is sprawled across the bed, arms flung out, snoring.

I'm not exactly sure when this sleep pattern starts. One day your wide-eyed children pounce on your head before the sun comes up; the next day you've had breakfast, lunch, and mowed the lawn before they crawl from their darkened caves.

Last year my eldest daughter, Hayley, then fourteen, was deep into her metamorphosis. Every morning we had the same routine. First, we'd ask Hayley to get up. Then, fifteen minutes later, we'd yell at Hayley to get up. Then, as school buses approached, we would go into Hayley's room and tip her mattress into the air so she slid, in a clump, onto the floor.

As actor Robin Williams might say: "Good Morning, Adolescent!"

THE CHEMISTRY OF SLEEP

Is this my fault? In a way, I might be tempted to argue yes. As working parents, my husband and I often got home late and, wishing to see the kids, let them stay up later than we should have. As they entered puberty, they drifted into staying up even later.

One mother of a sixteen-year-old boy says she and her husband regularly go to bed hours before her son, who, she says, refuses to go to bed before one or two A.M. "I don't even know what he's doing, some homework, maybe just wandering around and eating," she told me. "We've begged him to stop this because he's always exhausted, always sleep deprived. In the mornings we have to drag him out of bed, then he takes a half hour shower to wake himself up. And on the weekends, you can hardly get him out of bed at all. Last Saturday, he slept until three P.M.

Imagine, three P.M. It's like he gets up when the sun is going down."

This mother, too, blames herself. She says she let her son get into a bad habit of staying up late because after-school sports took up his afternoons, and he had to stay up later and later to finish homework. "When I was in high school, I still had a bedtime; I think I had to be in bed by ten P.M. no matter what," the mother said. "Maybe we should go back to that idea."

But Mary Carskadon, a sleep researcher at Brown University, says parental slack is only a small part of the story. Teenagers, she says, have a natural tendency to stay up later and sleep later. And for this, too, we can blame the brain.

Over the past several years, Carskadon has done a series of studies showing that teenagers stay up later and sleep later in part because of basic biology. It's not only because their bodies are growing, it's because, as they enter adolescence, their brains are changing in fundamental ways, as well.

In dozens of studies, Carskadon and her colleagues have found that teenagers start to secrete melatonin up to two hours later than when they were younger. Melatonin, sold in drugstores to treat jet lag, is one of the brain's sleep chemicals. As the day grows darker, it's secreted by the pineal gland and it helps make us drowsy.

Carskadon says as children enter adolescence they develop a "phase delay." They naturally start staying up later and sleeping later in part because melatonin flows into their brains later—most often around 10:30—and also lingers later in the morning, perhaps making them sleep later.

No one was more surprised by her findings than Carskadon herself. As a young graduate student at Stanford University, she once ran a camp to research teenage sleep patterns thinking she'd find that adolescents need the same amount of sleep as adults, around seven and a half or eight hours a night. But that's not what happened. In a controlled lab setting, Carskadon found that teenagers happily slept on and on, a bit over nine hours. And even then, they were sleepy in the middle of the day. A teenager's sleep need, in fact, far exceeds that of adults.

"We gave them a ten-hour window for sleep and we still had to throw them out of bed in the morning," Carskadon says.

Later, in experiments in her own sleep lab at Brown, Carskadon discovered the shift in sleep timing that seemed to emerge at puberty. She found that within a group of sixth graders, those further along in puberty naturally went to bed later. By measuring levels of melatonin in saliva, she also found that, as adolescence progressed, teenagers secreted melatonin later and later in the night.

"Before this it was assumed, and I thought, too, that the reason older kids went to bed later was only because of the psychosocial stuff, the friends, other things to do," says Carskadon. "It was a surprise."

Carskadon is fairly certain, given the timing, that such sleep shifts are linked both to puberty and changing brain chemistry. Some studies have suggested that melatonin levels decrease as puberty hormones rise, in particular luteinizing hormone, which helps spark ovulation and sperm production.

Carskadon is also exploring another intriguing idea. It could be, she says, that the release of melatonin became delayed because adolescents needed to stay up later for the group's survival. "Maybe, at one point in our history, it was important for young people, with good vision and strength, to be more awake and alert later in the day to protect the tribe," she says. "Something is going on that makes adolescents sleep differently from younger kids or older adults."

ALL of that could be just fine, of course. Pushed by biology, teenagers could stay up later, finish their homework, chat with their entire e-mail buddy list, and patrol the house protecting the tribe, all before going to bed at the appropriate melatonin hour.

But as it turns out, the grown-ups have screwed this up, too. Not only are we much worse than our parents at making kids go to bed at a reasonable hour, we've also decided to make high schools start earlier. Faced with increasing numbers of kids, rising costs, and competing bus schedules, high schools around the country, and to a certain extent around the world, have begun opening while the streetlights are still on. Some start as early as 7:10 in the morning, with students, after slogging through advanced placement calculus and physics, reporting to the cafeteria for their spaghetti lunches at 8:45 A.M.

That's left us with a disconnect between the amount of sleep teenagers need and the amount they're getting, a nation of sleep-deprived—and downright grouchy—teenagers. In fact, many adolescent researchers say that when a teenager seems out of sorts, touchy, and testy, one of the first things we should think about is whether they're getting enough sleep.

Amy Wolfson, a psychologist at Holy Cross in Worcester, who has collaborated with Carskadon on a number of studies, once surveyed three thousand high school students in Rhode Island and found that many were getting about six hours of sleep, far less than the nine hours they generally need. Other studies by Wolfson and Carskadon found that those teenagers who get less than nine hours of sleep, when given the opportunity to sleep in midmorning, tend to fall straight into REM sleep, a sign of severe sleep deprivation. What's more, teenagers who don't get adequate sleep do less well in school and score higher on batteries of tests measuring levels of sadness or hopelessness. In short, they feel crummy.

"It's scary how little sleep they get," says Wolfson, who's trying to design school health programs so that students—and their parents—are taught the importance of sleep.

"We're producing a group of students with sleep disorders," says Carskadon.

DEPRIVATION AND EMOTION

Ron Dahl at the University of Pittsburgh, another of the country's premier adolescent sleep researchers, has found that a lack of sleep can affect teenagers in a wide range of areas, including emotions. While sleepy teenagers don't experience different emotions from others, Dahl says they tend to have feelings that are less controlled and more exaggerated.

"It's not just that they get more negative moods, they are likely to be more silly, too," he says. "If they're frustrated, they're more likely to show anger; if they are sad, they're more likely to cry. They're less able to rein in an emotion. The feelings are more raw."

In a series of unusual experiments, Dahl has found that too little sleep can, in particular, impair a teenager's ability to do two important things at once, such as thinking and curbing emotions. When sleepy teenagers simply had to remember a letter they'd been shown on a computer screen seconds before, they did fine. But when Dahl superimposed pictures of emotion-provoking objects such as puppies (good) or fly-covered food (bad) on the computer screens behind the letters, the sleepy teenagers could no longer remember as easily as a control group that got adequate sleep. The sleep-deprived teenagers could no longer process emotions and think effectively at the same time, which means either activity, controlling emotions or remembering, could be impaired.

Dahl also believes that, without sufficient sleep, the brain may not have enough downtime to "retune." Neurological systems may need time to disengage so they can connect better later, much as parts of an orchestra must tune up alone before they can successfully play as a full group. And if various brain parts are not connecting smoothly, problems occur, an emotion can spin out of control, mild irritation at a mother can become a snarled "*Leave me alone!*"

"Without sleep you can perform in narrow ways, but it comes with a cost of something else," Dahl says.

Eve Van Cauter, a sleep researcher at the University of Chicago, takes it a step further. In one study, a group of young men allowed only four hours of sleep a night for a period of time, showed evidence of a whole range of hormonal dysfunction, including elevated levels of the stress hormone cortisol and an impaired ability to process glucose, a condition that can contribute to obesity and type-2 diabetes, both of which are on the rise in this country, particularly with adolescents.

"We found that with sleep deprivation, whole systems get out of whack," she said. "And teenagers are the most sleep-deprived segment of the population."

THE fifteen-year-old boy regularly goes to sleep at four A.M. He's bright, but his grades are horrible, largely because, fast asleep, he misses his bus and rarely shows up at his Philadelphia school. When he does show up, he promptly falls asleep in class. Not long ago his desperate parents took him to a sleep clinic run by Jodi Mindell in Philadelphia. He was found to have the most common sleep disorder in adolescence, a kind of extended delayed sleep phase. What happens, Mindell says, is that on top of the natural biological shift that makes teenagers stay up later, some, perhaps natural "owls" to begin with, push the sleep envelope even further. And they get stuck. The kids don't have insomnia. When they go to bed at four A.M., they fall asleep right away. Instead, their natural sleep phase, perhaps in part because of brain changes in puberty, has gotten way out of kilter with the rest of the world. If the pattern persists into adulthood, adults can adjust their lives, working the night shift, for instance. But teenagers, not yet in charge of their schedules, get in trouble.

The disorder can be fixed, but in a rather bizarre way. Teenagers, who are often happy to hear this, are told to push their bedtimes even later. If they go to bed at four A.M., they are told to go to bed at six A.M. Then, each successive night, they are supposed to work themselves all the way around the clock until they go to bed at an acceptable bedtime. This takes a great deal of motivation and cooperation, not particularly strong suits for an average teenager. Many times, by the time the sleep-deprived teenagers show up at sleep clinics, they're depressed and their families are stressed and in disarray. It's hard to sort out what is leading to what. In the case of the fifteen-year-old, for instance, he refused to stick to his schedule. "He made it all the way around the clock and just kept going," says Mindell. "It takes a huge commitment from kids to fix this and it's hard. Frankly, those of us in the field groan when one of these kids walks in the door."

Judy Owens, a pediatrician who runs a sleep disorders clinic in Rhode Island, says that in addition to delayed phase sleep, other sleep disorders that are common with adults often begin in adolescence, including narcolepsy, in which people fall asleep at the drop of a hat at inappropriate times, and insomnia, in which people can't fall asleep at all. The onset of insomnia, Owens thinks, can be traced back to the ongoing development of the frontal lobe, which acts not only as the policeman but also as the worry machine in the brain.

"With brain development of the prefrontal cortex, teenagers are getting to be better planners, but they also get to be better worriers," she says. They plan ahead and worry about that test that's coming up tomorrow, all those things. You don't see that in younger kids."

That frontal lobe development may also influence dreams. Although research on this is not extensive, there's evidence that teenagers start to dream differently, too. In his book *Children's Dreaming and the Development of Consciousness*, longtime sleep researcher David Foulkes says that as the brain and cognition develop, children's dreams take on a more coherent narrative flow, are less about animals and, as a sense of self evolves in adolescence, increasingly tend to involve the dreamer as the main character.

Many sleep researchers acknowledge that teenagers are not helped by the 24/7 soup we live in. "Everything is twenty-four hours," says Judy Owens. "You want cartoons at three A.M.? You can have cartoons at three A.M."

Owens recently ran a sleep experiment at a summer camp for two hundred kids and found that their sleep schedules were erratic and they wanted help in fixing them. Many had televisions in their rooms, parents who worked long hours and stayed up late themselves, and offered little encouragement about the need for sleep. When they tried to catch up on sleep on weekends, their parents thought they were lazy.

"These kids said they had a lot to accomplish and they realize that they're not as motivated or good in school or in sports when they don't get enough sleep; they make the connection," Owens said. "But they also said they needed someone to help them organize their time. They aren't lazy; they're just sleeping far less than they should be."

AN ALICE-IN-WONDERLAND WORLD

Why do any of us—teenager or adult—sleep at all? It seems an obvious question, but, in fact, there's nothing obvious or simple about sleep. David Dinges, a sleep researcher at the University of Pennsylvania School of Medicine, says new tools have meant that we've recently found out a great deal about sleep. "It's astonishing what we know," he says. Still, he concedes, big black holes, "true mysteries," remain. And even the things we do know are odd at best. An Alice-in-Wonderland world where genes dubbed "clock" and "timeless" turn on and off, muscles become paralyzed and the brain decides, pretty much on its own, what it wants to do—the activity of sleep, says Dinges, just gets "curiouser and curiouser."

All animals sleep, in their fashion. Dolphins and whales sleep with half their brains so that the other half can move them to the surface for air. Some ducks do the same thing. When they sleep in a group, the duck on the outside sleeps with only half its brain so that it can keep one eye open, watching. Cows sleep standing up. Giraffes kneel, tucking their long necks around to rest on a back knee. Even fruit flies sleep. Give them caffeine and they'll stay hopping around later; interrupt their rest and the next day they'll stubbornly sit in a tiny clump longer, paying off an accumulated sleep debt like the best of the tenth graders.

General sleep patterns have emerged. Big animals sleep less than smaller animals with rapid metabolisms, lending credence to the theory that sleep helps animals restore energy reserves. Animals that can take care of themselves fairly well when they're born, dolphins for instance, spend less time in REM or, rapid eye movement sleep, after birth than more dependent animals. This adds weight to the theory that sleep, and perhaps REM sleep in particular, is important for the continued brain growth of the more dependent ones. Human babies, who spend 50 percent of their time in REM sleep, fall in the middle range.

One of the most intriguing findings recently involves REM sleep and learning. Several years ago there was an experiment showing that rats that learned a circular maze in the daytime, rewarded with chocolate sprinkles, rehearsed the same maze in their dreams at night. Researchers pinpointed the area of the brain that was active when the rats learned the maze and then watched the same area light up when the rats slept. Other recent studies have shown that the brain nerve fibers of kittens that slept grew more than those that didn't sleep.

Clif Saper, a sleep researcher and neurologist at Harvard and Beth Israel Deaconess Medical Center, on the other hand, thinks one main function of sleep is to give our nerve cells a rest, to "wipe the blackboard clean and make changes in synapses so that they can take information in again the next day."

"The nerve cells get a lot of biochemical signals all day; they may need to take a break and sort through all that," explains Saper. It may be a way for nerve cells to sort through the junk mail, keeping the good stuff, throwing out the trash.

Eve Van Cauter believes that, in the end, we must sleep for a host of reasons that are crucial for survival. Rats, she points out, die faster without sleep than without food, possibly because their immune systems fail. One of the signs that a sleep-deprived rat is about to die is that it gets sores all over its long, thin tail. "The point is, we need sleep for everything," says Van Cauter.

"It may be," adds Dinges, "that sleep was organized for one set of functions and has taken over other functions."

NIGHTTIME UPS AND DOWNS

Plotted on a graph, sleep looks like a small roller coaster. We float down in stages into deep slow wave sleep, then move back up periodically into the more active dream-inducing REM sleep and then back through the various stages of deep sleep again, roughly in ninety-minute cycles.

Stage one sleep is light, our brain waves are still active, we're easily awakened by a knock on the door. During stage two, body temperature drops and brain waves begin to slow. Stages three and four are the deepest sleep. Finally, we flip into REM sleep, often referred to as paradoxical sleep because brain waves are nearly as active as when we're awake. During REM, muscle tone in limbs is lost, so while most of our vivid dreams occur in REM, we cannot act them out because our muscles won't work. (If you cut brain connections that bring on that muscle paralysis, a cat will chase pretend mice in its sleep and people, with certain sleep dysregulations in this area, will try to act out their dreams, trying to hit all those mythical home runs, for instance.)

Generally, we follow a twenty-four-hour cycle, a time frame probably set when amoebas reacted to the turn of the earth on its axis. Studies of those who've volunteered to live in caves or dark labs, though, find that most will push the cycle a bit out of sync, to about twenty-five hours, which means they slowly drift out of phase with day and night.

To keep from drifting out of phase, our internal biological clock takes hints from the environment, in particular from levels of light. Light, passing through photoreceptors in the eye's retina, hits two pinhead-sized clusters of cells in the hypothalamus (called the suprachiasmatic nucleus, or master clock), releasing hormones that make us sleepy, including melatonin.

As humans, we use a grab bag of other tricks as well—regular mealtimes, TV shows, multiple alarms, Starbucks—to keep us on whatever schedule seems necessary.

The timing system, however, is just one part of sleep. While it may tell the brain *when* to sleep, it takes another whole system to make the brain sleep, and to wake it up. No one is exactly sure how it works, but it appears that one set of neurotransmitters keeps the brain alert and another puts it to sleep, with both triggered by separate parts of a busy hypothalamus. An area that produces a chemical called hypocretin, some scientists say, helps keep the brain awake. Researchers recently found that humans with narcolepsy lack or have very low levels of hypocretin, perhaps destroyed by an autoimmune reaction. Another segment of the hypothalamus, the preoptic area, may put the brain back into sleep. According to Clif Saper, when sleeping sickness hit in 1920, those with damage to the preoptic area became insomniacs.

Both regions of the hypothalamus, the awake center and the sleep center, get clues from another cluster of cells in the hypothalamus, called the master clock, as well as dozens of chemicals that build up in the brain while we are awake. One of those is adenosine, a by-product of brain-cell activity that increases and, eventually, acts on the preoptic area to make us drowsy. Whether we know it or not, all of us are adept at fiddling with this particular brain chemical. Sleep-inducing adenosine is blocked by the caffeine in a cup of coffee.

AT Washington State University, neuroscientist Jim Krueger is trying to understand sleep by breaking it into its smallest components. In his lab, he is trying to coax neurons to sleep.

In fact, with single neurons, Krueger says, it's impossible to tell whether they are awake or asleep. Some brain cells fire faster when they're promoting sleep than when they're not. When I spoke with him, Krueger was adding neurons to his little petri dish bed one by one until they formed a "neuronal group," large enough to fall into a concrete cyclical pattern that we'd describe as sleeping and waking.

Deep down, Krueger thinks we are more like dolphins than we might think. Parts of our brains fall into a deeper sleep than others, he says, largely because they've been used more during the day. Of course, unlike dolphins that sleep with half their brains, eventually the whole human brain falls asleep. We don't do the dishes with half of our brain while the other takes a nap. But some studies have shown that parts of the human brain seem to sleep more than others. For example, several years ago researchers in Zurich had a group of people repeatedly move their left hands and later watched their brains as they fell asleep. The area of the brain devoted to the left hand fell into a deeper sleep than areas devoted to the right hand.

Krueger believes the brain is arranged this way because it's selfish. It rests the parts it has used the most and, at the same time, strengthens the connections of those same parts because they seem to be important to us. As it happens, Krueger says, the same fifty brain chemicals that make our brain get drowsy also help it build synapses between brain cells. That means that areas of the brain that are most used will build up the most sleep-promoting chemicals. And those same chemicals then construct and strengthen brain connections while we sleep.

This way, he says, the brain, which really must have some good reason to sleep because it's such a long period of time when it's "not vigilant against predators, not eating and not reproducing," could preserve needed lessons and take a break, too. His idea is that when enough neural groups get used and build up enough of those chemicals, they put that local area of the brain to sleep. Then when enough local areas are asleep, something happens to make the whole brain fall asleep.

"You know," he said, referring to napping areas of the brain, "neurologists have said for years that there are people who can be simultaneously awake and asleep."

What does all that mean for teenagers, who often seem to be simultaneously awake and asleep themselves? For one thing, during adolescence, the brain's sleep systems are still developing and shifting around. Clif Saper, the Harvard sleep researcher, points out that the preoptic area of the hypothalamus, which he believes helps put us to sleep, shrinks as we age. That may be one reason why older people have trouble reaching deep sleep. But in teenagers, he says, that sleep-promoting area is well endowed.

Other shifts are also taking place. Over the course of adolescence, slow wave sleep, the deepest, declines by as much as 40 percent. That's why doctors, faced with a young child who walks or talks in her sleep or wets his bed, tell parents not to worry. Those activities happen during slow wave sleep and, as children enter adolescence and slow wave sleep declines, those actions often disappear.

Mary Carskadon believes the decline in slow wave sleep may come about because of the changes in the teenage brain as it develops. Slow wave sleep needs a certain density of neurons in the cortex to occur. And the teenage brain, as we've seen, is slashing and burning its cortical gray matter like mad.

"Adolescents," Carskadon says, "are different in how they sleep."

THE sun was just peeking over a nearby hill when Kelly Crossett flung her backpack over her shoulder and headed out the door to school. It was seven A.M., on a backcountry road of Katonah, a small town north of New York City, and Kelly was the only one in sight moving around.

At seventeen, Kelly was just starting her senior year at John Jay High School. She was a star student, president of the student government, editor of the yearbook, captain of the volleyball team. As she left the house that morning, her long hair was carefully combed, her jeans neat. She looked wide awake. Yet the night before, like most of her nights, she'd gotten only five hours of sleep.

"I guess I'm tired all the time," she told me as we drove to her high school campus.

But several years ago it was even worse. In those days, Kelly's school started at 7:10. That meant that Kelly and other students often had lunch periods as early as 8:45 in the morning. Macaroni and cheese, pizza, chips.

"It was pretty ridiculous," concedes Kelly. "Mostly I didn't eat anything."

It was so ridiculous that parents in Katonah protested and, citing Mary Carskadon's research into biological sleep shifts as evidence, persuaded school officials to start high school classes a half-hour later. Other school systems, such as those in Minneapolis, that have taken Carskadon's research to heart and delayed high school starting times even later, have noticed improved overall mood and attendance, although it can create problems with scheduling after-school activities. But since adequate sleep has been concretely linked—in science labs, schools, and homes—to a distinctly friendlier and more functional teenager, shouldn't sleep deprivation be one of the first things we consider when faced with a distinctly crabby and dysfunctional teenager?

In Katonah, Kelly said the time shift had helped, particularly initially. But because many activities have now been moved to the mornings, and because traffic has gotten so bad, students still find they have to get to school as the sun comes up.

As she pulled into the school parking lot on the day I accompanied her, there were dozens of students, dressed in jeans and big jackets to ward off the late autumn chill, sipping coffee.

I asked one, Adam, sixteen, who wore a red baseball cap, a ski jacket, and held an enormous stainless-steel cup of coffee, how much sleep he'd gotten the night before.

"Oh, it was pretty typical," he said. He'd had sports after school, piles of homework, went to bed about two A.M. and got up before seven for school—about five hours of sleep.

Does he ever fall sleep in class?

"Oh yeah," he answered, as if this were a silly question, "especially in science. All the kids fall asleep. But the teachers are really nice; they don't wake us up."

Chapter 12

Falling Off the Tracks

New Dangers from Old Devils

With her blond hair pulled neatly back in a bun, wearing a turquoise turtleneck and starched white pants, eighteen-year-old Michelle hardly looked like a heavy drinker. But every weekend for the past two years, she said, in the soft sweet voice of the pastor's daughter that she is, she set out "to get as drunk as possible."

On one Friday night, typical of most, she drank four tall glasses of vodka and orange juice, three glasses of rum and Coke, and a half-bottle of coconut rum. Michelle then threw up all night and, the next day, wasn't sure where she'd been.

On Saturday night, she followed up with two forty-four-ounce glasses, "the big tall ones you get at gas stations" of vodka mixed with Mountain Dew. She was at a beach party, but all she could really remember was sitting on a car bumper in a parking lot, vomiting.

"I don't know why I did it," said Michelle. "It wasn't peer pressure . . . a lot of my friends don't drink at all. My parents don't drink. Once at a friend's house, their parents gave me a small glass of red wine and I just decided not to stop. I drank every weekend night. Mostly, I drank hard liquor. Maybe I was bored. It was a small town and it seemed like there wasn't anything else to do."

In the middle of her drinking years, Michelle heard about a brain-scanning project at UC San Diego and, out of curiosity and for the $100 it offered, decided to volunteer. The experiment was designed to test the cognitive abilities of teenagers who drank heavily. As a subject, Michelle climbed into a brain-scanner and, looking at a small screen inside the tube, was shown lists of disconnected words and asked to remember them: jeans, balloons, apples. Then she had to follow the movements of a squiggly line and a little dot that darted here and there.

"I did the best I could. I hope I'm OK," Michelle told me, worry in her voice. She clutched a bottle of water, her drink since she had given up binge drinking several months before. "My reflexes are still good; I still get good grades. I hope I haven't hurt my brain."

THE findings about what's happening within Michelle's brain are not yet in; the research is ongoing. But there are hints that at least some alcohol-soaked teenage brains like Michelle's are not doing as well as we might hope. It's well known that hard-core drugs like cocaine, heroin, or speed can affect the brain, inducing addiction in young and old. Evidence suggests that heavy use of one of the most popular new drugs, Ecstasy, in particular, may cause severe damage to brain cells, specifically those that produce dopamine and serotonin.

But alcohol has been considered a lesser evil, at least in teenagers. Adolescent brains have long been thought of as more resilient, able to resuscitate synapses, even after a good sousing.

Now, however, there are strong hints that that might not be the case. Sandra Brown and her colleagues Susan Tapert and Greg Brown at UC San Diego are conducting a series of long-term studies of heavy and binge-drinking teenagers like Michelle. The early results show that alcohol may be even worse for the brains of teenagers than it is for adults.

One study found that teenagers who drank excessively—they averaged two drinks a day for two years—consistently recalled 10 percent less on memory tests than nondrinking teens, a rate that was also worse than that of adults with a history of alcoholism. In some cases, the impact was seen years later, too, after they'd stopped drinking for months. A memory loss of 10 percent may not, on the face of it, seem huge, but as Brown says, "it can mean the difference between an A and an F."

And additional studies by Brown have confirmed the point. One showed that older adolescent girls, all binge drinkers, had less overall activity in brain areas used for certain cognitive tasks, such as finding your way to the store and remembering what you went to the store for. In fact, those areas of the brain, including crucial cortical areas of the frontal and parietal lobes, were less active even while their brains were at rest.

"We now think that the teenage brain may be more sensitive to alcohol than we thought," said Brown. "There's a lot of development going on in the brain during adolescence. And we know that at other times of major development, in utero for instance, the brain is quite sensitive to neurotoxins like alcohol."

Teenagers, of course, seem determined to test Brown's hypothesis. In my own neighborhood, the stories are legion. Just last fall, all school-related dances were banned in the upscale suburb of Scarsdale, New York, after five teenagers—two boys and three girls—were rushed to the hospital after collapsing from drinking multiple glasses of vodka and orange juice before a homecoming dance. One sixteen-year-old girl, nearly unconscious when she reached Scarsdale High School, had to have her stomach pumped. Police estimated that as many as two hundred kids at the dance, including boys and girls as young as fourteen, were severely drunk, with some vomiting in garbage cans and others either unconscious or incoherent and on the verge of passing out. Some of the kids used vodka sneaked from their parents' liquor cabinets and disguised in Poland Spring water bottles. Last year, a popular high school athlete in Harrison, a suburb near Scarsdale, died after he was punched and fell and hit his head at a drinking party, and in Rye, New York, last spring a number of students at a dance, suffering from acute alcohol intoxication, had to be hospitalized.

According to federal surveys, 30 percent of twelfth graders, 26 percent of tenth graders, and 14 percent of eighth graders reported heavy drinking—that is, like Michelle, they had at least five drinks in a row at least once in the previous two weeks. Alcohol remains the most commonly used psychoactive substance during adolescence, and binge drinking is much more common among white and Hispanic high school students than blacks. Other studies show that children who start drinking before age fifteen are five times more likely to be heavy drinkers as adults, ten times more likely to be involved in a fight, seven times likelier to be involved in a car accident, and twelve times more likely to be injured.

Most experts say solutions to excessive teenage drinking are far from easy. On the one hand, they don't want to exaggerate the evils of alcohol, but they feel its dangers need to be emphasized more. While many parents remember drinking in high school, those who study teen drinking today say the patterns have changed, with bigger parties and considerably more binge drinking—with the goal being total oblivion—by a broader spectrum of kids, including younger kids and more girls. After yet another fighting and drinking incident in Harrison involving some members of the high school football team, the coach pinned the blame on the kids whose judgment, he said, was "very, very disappointing." Other experts point to the parents, who, they say, are often lenient about alcohol because they're so happy their children are not involved with marijuana or cocaine. Michael Nerney, an expert on preventing substance abuse, says parents contribute to the problem by not sending clear messages that drinking is not a rite of passage and not effectively monitoring their teenagers. After the drinking incident in Scarsdale, the frustrated high school principal, John Klemme, pleaded for more "abundant parental supervision."

"Alcohol may be more harmful to teenagers than is understood by the general population," says researcher Sandra Brown. "We think of alcohol and even nicotine as relatively benign drugs for teenagers in terms of adverse impacts on cognitive function. We need to take binge drinking more seriously. It should be a red flag for parents."

DAMAGE TO THE HIPPOCAMPUS

It's a flag that Scott Swartzwelder at Duke University has been waving for some time. Seven years ago, Swartzwelder and his colleagues published a study that hinted at potential brain problems with excessive alcohol, at least in adolescent rats. Studying the rats' hippocampi, he found that it took the equivalent of just about two beers to decrease memory-related functions in adolescent rats, while it took twice as much alcohol to cause similar damage in adult brains.

"We got into it because we realized that no one had studied it," said Swartzwelder, who said his interest was further spurred because his father was an alcoholic. "Adolescence is when most people start drinking and often do the heaviest drinking of their lives, but the whole area had been ignored."

Alcohol, Swartzwelder says, is now thought to affect memory by interfering with the neurotransmitter glutamate, which carries excitatory, wake-up messages from one neuron to another.

There is still a great deal that is unknown about memory and learning in the brain, but generally it's thought to go something like this: In their normal state, brain cells exhibit a general hum of low-level activity, even when the brain seems to be doing little at all. But that hum reaches a higher level of gossip when the brain is stimulated by something, say the sight of a girl's face. That stimulating message is passed along, in part, by glutamate released from one neuron to glutamate receptors on a neighboring neuron. The glutamate opens channels on the receiving neuron, creates an electrical charge, and passes the image of the face along.

But suppose you meet someone whose face you want to remember for a long time—a face you might want to have dinner with Friday night. In that case, brain cells, in particular those in the hippocampus, start firing in a stronger rhythmic way, in effect shouting "pay attention to this!" If a strong enough pattern develops in two neurons at the same time, glutamate released from one neuron will not only operate its regular receptors on the next neuron, but also special receptors—NMDA receptors—that are normally closed tight with magnesium.

And as the magnesium floats away, the special NMDA receptors in the receiving neuron also open up and permit calcium ions to flow into the cell, creating what Swartzwelder called a "cascade of biochemical events" that somehow—no one is quite sure how—change that target neuron to the point where regular messages to regular glutamate receptors in the future are more easily recognized and passed along. In this way, the image of a pretty face would have burned an enduring pattern into your neural circuits that's unlikely to be forgotten, at least through the weekend, a process that's thought to be a foundation for learning and memory.

It's not precisely known how alcohol interferes with this intricate memory process, although some think excessive drinking may interfere with the flow of magnesium, keeping it from being carried away and, as a result, leaving the NMDA memory gates closed.

What is known, as far as Swartzwelder is concerned, is that the whole process seems to have more of an impact in adolescents who drink. Studies from Swartzwelder's lab show that drunk adolescent rats are much worse at learning mazes if they have alcohol in their systems than adult rats with the same amount of alcohol. Even if the adolescent rats drink heavily while young but are allowed to grow to maturity without alcohol, they exhibit problems. The early drinkers, after they grew up, learned mazes as well as rats that did not drink while young, but they flunked the tests if they had the equivalent of just one drink, perhaps because their brains were compromised early on and they could not function as well under duress. Surprisingly, Swartzwelder has found that alcohol is much less sedating in adolescents than in adults, which may seem like a good thing but isn't. "If teenagers don't feel sleepy after they drink, they may think it's OK to get behind the wheel or climb out on that ledge," he said. And Swartzwelder has also found that a single dose of alcohol impaired learning ability much more powerfully in younger people, aged twenty-one to twenty-four, than it did in those aged twenty-five to thirty, even after just enough alcohol to meet the minimum definition of being legally drunk.

"We now know that the teenage brain is different, and one way it's different is that it seems to be more sensitive to alcohol," said Swartzwelder. "The brain is still developing and the things you do can mean a lot. Once adolescence is gone it's gone for good. What you do then potentially can have an impact for the rest of your life."

ALCOHOL AND CELLULAR SUICIDE

Some scientists believe that it's when the heavy drinking stops that a good part of the damage may occur. When the brain is alcohol-soaked and NMDA-glutamate receptors are blocked, it reacts by creating more sensitive receptors. Then, once the drinking has stopped, there may be too many receptors in the brain that let in too much calcium. And while a little calcium can be a good thing and help promote memories, according to Mark Prendergast, a neuroscientist at the University of Kentucky, "too much of it is not good at all." An overload of calcium can turn on "suicide" genes in a cell, a mechanism that's similar to what happens in some strokes.

In his study, Prendergast and his colleagues first doused the slices of rat hippocampus with alcohol, then withdrew the alcohol and monitored calcium uptake and cell death. So far, preliminary results show there is, indeed, an overabundance of calcium being admitted into cells and a "dramatic cell death" in the hippocampus after the alcohol was withdrawn. Early results also suggest that the damage is much worse in the brain slices from the adolescent rats. It's a lesson that Prendergast believes applies to human teenagers, too.

"The big problem comes when you come back home after a week of vacation and partying and the brain is trying to compensate for all the alcohol it had," said Prendergast. "That's when you get small seizures, some of them overt, and anxiety, depression. And it may be because of the cell death in the hippocampus."

Whether that damage is reversible is an open question. There are hints from a number of early studies that it may not be, at least not completely. Those who have neurological and motor damage from drinking later in life often have a history of early binge drinking. One brain-imaging study at the University of Pittsburgh found that the hippocampi of heavy teen drinkers were 10 percent smaller than nondrinkers. It could be that the brain structure they were born with initially led them to drink. But Prendergast and others say the new evidence suggests "drinking very well might be a reason" for the damage.

IN San Diego, Michelle was talking about alcohol with her friend, Keith, who had come with her to the university where she had her brain scans. They were trying to figure out why it was that some teenagers, including Michelle, get caught up in drinking while others don't.

Keith, for instance, said that he had never really gotten drunk—even though he had, on a few occasions, tried. An eighteen-year-old with the build of a football player and bright red hair, Keith told the story of a party in the California desert when he drank "a half bottle of vodka, several shots of tequila, a wine cooler, and a piña colada," and "there was nothing; no buzz whatsoever."

He said he had no idea why he did not get drunk while others, with just a few drinks, seemed to float into alcoholic oblivion. What was it for the others? Was it bad luck? Bad genes?

In their own lives, they said, they see evidence for both. They both talked about their friend Nicole, who, Michelle said, was from a "long line of alcoholics." For many years, Nicole had been Michelle's drinking buddy. But while Michelle had sobered up and was on her way to college, Nicole had stayed at home, working at a pizza parlor, using her money to "get drunk every night on ninety-nine-cent beers."

"I think her whole family is going to turn out that way," said Michelle.

But Michelle and Keith also spoke of another friend, whose family—parents and some of his brothers and sisters—were also alcoholics. Yet this boy had, early on, decided he wanted none of it. He never drank, studied hard, and was on his way to Purdue University.

"So that makes me wonder, how can it be genetic?" asked Keith. "It's hard to say. I see it going all sorts of different ways even with the kids I know. Some get into it and can stop and others can't. It's weird."

No one is more familiar with that weirdness than John Crabbe, a neuropsychologist at the Oregon Health and Science University, who also recently reviewed all the literature on genes and alcohol, a picture, he says, that is muddy at best. There's no question that alcoholism runs in families. Studies involving twin children of alcoholics who were raised in different homes tracked the children to see if they became drinkers. The results repeatedly found that 50 percent of the risk of alcoholism can be traced to genes, a number similar to risk levels associated with other substance abuse problems. But that still leaves a lot of room for the influence of environment.

"There is growing literature that says that genes might predispose a person to drugs of abuse, including alcohol," said Crabbe. "But even [among] those of us who are firmly committed to a medical model and those who study human genetics and have a lot of reason to be biased, we all know that the nongenetic contribution is equally important."

Because alcoholism is such a persistent problem, scientists have continued to try to locate the genes responsible for the genetic part of the equation. A few years ago, there was a flurry of excitement when researchers thought they'd found the "the alcoholism gene."

Today that excitement has waned. More recent studies have found that the gene, involved with levels of dopamine, is not consistently found in alcoholics. For the most part, researchers have targeted genes that play a role in the brain's reward system, the genes linked to neurotransmitters that make us feel good, in particular dopamine, serotonin, and the endorphins. Alcohol is known to act on many of those neurotransmitters, making addicts want more alcohol to get more of the feel-good chemicals in their brains. Although there are hints of associations here and there, and hope that the recent mapping of the human genome will help, Crabbe said no one yet has hit genetic pay dirt.

Alcoholism, like other complex diseases, is likely to be associated with many different genes. Efforts are under way to try to break down the components of alcoholism, withdrawal, and desire, for instance, and find genetic links to those segments of the behavior. Crabbe thinks that approach might work eventually, but at the moment, "it's a tough game, not hopeless, but hard."

So what would Crabbe tell a fourteen-year-old boy whose parents were both alcoholics?

"I'd tell him he had an eightfold higher risk and to watch it," said Crabbe. "But I would also have to tell him that he might have no risk at all. Of all the genes involved in alcoholism, maybe he got none. Maybe he was lucky."

WHEN I talked to a group of kids at Phoenix House, a residential drug rehabilitation center, most had been heavy drinkers, along with using an assortment of drugs. What had happened to them, I wondered? Something had set them on that path. Perhaps an inborn need to be on the edge, perhaps an environment that made it possible, if not imperative, perhaps both.

For John, a pudgy boy with a crew cut and urgent manner, the need was pushed by anger and family history. His parents were divorced (his father had been a cocaine addict). His mother worked long hours as a nurse. He was close to his grandfather, who lived with him and took him to baseball games. But when John was fourteen, his grandfather died and John was a mess. "I was angry that he was taken away. I didn't understand. I was lonely. I was a latchkey kid. I just started doing stuff because I was so angry, I think."

But other kids get angry and stressed out. Why did he have to go so far? Why did he feel a need to become not only a drinker and a drug user, but also a drug dealer, peddling Ecstasy to neighborhood children? John looked down and shook his head. He didn't know.

For Julia, a pretty girl of seventeen, the choices were more predictable. Her house on Long Island was chaotic; her mother was a drug addict and at one point she'd been sexually abused by a friend of her mother's. One afternoon, she found some marijuana in a drawer; she tried it and did not look back. She became a chronic runaway and she drank constantly; she skipped school, even though she loved basketball and was a good student. Looking back, she says she thinks she was born the way she was to a certain extent. As a child she would climb the highest trees she could find and hang on to the back of trucks with her skates to get a ride. Was that because she was trying to forget what was going on in her house? Was that related to her genes?

"I don't know why I did it," she said. "I think I am a person who needs rules. When I was at my father's house there were rules that were enforced and I felt better; I had to go to school and to church. But then I would get stressed and get impulsive."

SMOKING AND PANIC

The young girl started smoking when she was fourteen and quickly moved on to a pack a day. When she was twenty-two, she started having panic attacks. Out of nowhere, she would suddenly feel as if she were going to die; her heart would race, her palms would sweat, she had trouble breathing.

Most would probably guess that the girl's smoking was somehow related to her anxiety. She was a worrier so she smoked; she was a worrier so she had panic attacks. But two years ago, a long-term study found just the opposite. Researchers at Columbia University and the New York State Psychiatric Institute found that teenagers who were heavy smokers had a risk of developing panic attacks later in life that was fifteen times greater than non-smoking adolescents, even though their anxiety levels were no higher than others to begin with.

The study of about seven hundred youths was just an association study, a study in which scientists find a suspicious link between two things. But it's based on some solid theories. It's long been known, for instance, that panic attacks can be triggered by respiratory problems. After a few years of smoking, lung capacity is reduced, smokers take in less oxygen, and exhale less carbon dioxide. Carbon dioxide in the blood stimulates breathing and too much of it sends a signal to the brain that it's suffocating, setting off a false alarm and a panic attack.

"Most think that people smoke because they are anxious," said Jeffrey Johnson, a researcher on the study, published in the *Journal of the American Medical Association* in November 2000. "But we found it was just the reverse. Those who smoked in adolescence had an increased risk of panic attacks; somehow it changed something in their brains and they were more susceptible later on."

CAN smoking in adolescence, like alcohol, permanently change the brain?

A few years ago no one would have even asked that question. As Theodore Slotkin, a longtime nicotine researcher at Duke University says: If you check the scientific literature for anything about teenage brains and the effects of nicotine "from about 5000 B.C. to 1999, there would be zero."

That's because, as in other areas, most scientists believed that the structure of the brain was essentially finished long before adolescence, and any damage in those years would be minimal. Over the past few years, though, Slotkin and a small number of other researchers have begun to find evidence to the contrary. In fact, damage to the adolescent brain from nicotine, in many ways, mirrors that from alcohol. Researchers are now finding that it may cause greater damage to brains that are less mature.

Slotkin, who spent twenty years researching the impact of maternal smoking on fetuses, has now turned his attention to adolescents and smoking. He's been astonished at what he and others are finding. "I knew that the brain was developing when it was a fetus and you get these problems. Now that we are cognizant that the brain is still developing in adolescence, I began to wonder if we are still getting permanent changes then, too," Slotkin said. "And the answer is yes."

According to Slotkin, it's amazing that no one has looked at this question before, since studies have consistently shown that if you start smoking in adolescence you become more addicted than if you start later and "the majority of people start smoking in adolescence." Thinking that there was little real damage because the brain was already organized, he said, was "a really dumb concept to begin with when you think of it." Each day, he adds, three thousand American children under the age of eighteen begin smoking.

NICOTINE works in the brain in several tricky ways. It mimics the actions of acetylcholine, one of the most widespread neurotransmitters in the brain, but targets only one set of acetylcholine receptors. Nicotine does much of its work on the presynaptic cell—the neuron that is releasing the neurotransmitter rather than receiving it. If you have a dopamine-producing neuron with nicotine receptors, nicotine can stimulate those receptors and cause more dopamine to be released into the brain.

Nicotine's reach is broad. At least twenty different neurotransmitters are influenced by nicotine. It's been shown, for example, to increase levels of the excitatory neurotransmitter glutamate, igniting communications between brain cells and actually enhancing memory and learning. In this way, it's been shown to be helpful for kids with attention deficit and even older people with Alzheimer's.

Researchers are particularly interested in how nicotine acts in a midbrain area—the ventral tegmental area—which is rich in neurons that produce dopamine, one of the main chemicals involved in the reward systems of the brain, which makes us feel awake and alive (and that's so pleasing to rats and cocaine addicts that they will self-stimulate this area rather than eat or have sex).

How this works in adolescents, said Slotkin, is "a bit shocking." In one study, Slotkin found that if he gave rats doses of nicotine that simulated the levels achieved in typical smokers, the adolescent rats produced twice as many nicotine receptors in the ventral tegmental area as adult rats. And the increased numbers of receptors, which lead to increased cravings for nicotine, lasted for at least a month after nicotine treatment was stopped, a long time in an animal that only lives two years.

Other studies by Slotkin and his colleagues have found that female adolescent rats, given nicotine, experienced damage to the brain cells in their hippocampus, and, said Slotkin, "no one voluntarily gives up 10 percent of an area that's involved in learning and memory." Such damage may be restricted to females because estrogen is known to produce increased hippocampal cells at certain times of the estrous cycle and those new cells may be more vulnerable to damage. The loss, Slotkin said, shows up only in adolescent female rats, which may mean that teenage girls are particularly susceptible to the impact of nicotine on their brains. Other studies have found that adolescent rats exposed to nicotine had, for reasons that are unclear, lower serotonin levels, one marker for a heightened risk of depression. Their immune systems, too, measured on a cellular level, were only half as effective as what they should have been.

Slotkin feels his findings potentially have great implications for human teenagers, in part because they support what is already known about the behavior of teenage smokers. Smokers are known to be more prone to infections and depression. And, Slotkin said, those changes might very well come about because of miswiring that occurs in the adolescent brain when it is "insulted" by nicotine. Indeed, following up on these findings, Joseph DiFranza and his colleagues at the University of Massachusetts Medical Center found that teenage girls get hooked on smoking much more easily than boys. "Adolescence is a time when the brain cells are still developing and getting wired up and neurotransmitters are changing. It might be that nicotine can make a sizable change to wiring at that point," Slotkin said.

What's more, he said, at least in rats, the immune system response was found even after the rats had grown up and had stopped getting nicotine for a long period of time. Another study by a colleague Ed Levin found that teenage rats learn to self-administer nicotine faster than adults, and those that start in adolescence give themselves twice as much nicotine as grown-ups as do rats that are first exposed to nicotine as adults. "It looks like nicotine has some effect on the final sculpting of the brain as it prepares for adulthood," said Levin.

The fact that science has not found this before does not surprise Levin. Since most assumed that teenagers were similar to adults, they have been long ignored by researchers. No one yet knows the long-term impact of antidepressants, now regularly prescribed to teenagers, or even what impact there will be on the girl who starts taking estrogen in birth control pills at fourteen instead of twenty-five.

Different brains, of course, do not always mean worse brains. Levin's own studies have also shown that adolescent rats, if they get a Froot Loop reward, can learn mazes much faster than younger or older rats, a fact that also doesn't surprise him as he watches in awe as his fourteen-year-old daughter easily masters algebra. "Teenage brains are different; sometimes that's a good thing," Levin said.

And it isn't really a mystery to him or Slotkin why teenagers, even given what we now know about cigarettes, continue to smoke. A few years ago there was a Canadian study in which one-third of the cigarette packages had a skull and crossbones put on the outside.

"And just guess which package the teenagers wanted," said Slotkin. "You figure it out."

Chapter 13

Into Another World

When Things Go Wrong

THE disease, or diseases, that we call schizophrenia can occur almost anytime. One rare severe form starts in childhood. But John Forbes Nash Jr., the Nobel prize–winning mathematician who was the subject of the book and movie *A Beautiful Mind*—an odd, remote child—did not have a classic full-blown psychotic episode, the hallmark of schizophrenia, until he was 30 years old.

Yet, the majority—the vast majority—of schizophrenics get sick in adolescence.

And males are more likely to get the disease, and get it earlier, than females.

Why?

Depression follows a similar though slightly altered pattern. Children can be seriously depressed, as can old people, and levels peak in middle age. But depression also has firm roots in adolescence, with rates starting their most dramatic climb during the teenage years. And girls, starting at about age thirteen, get severely depressed far more frequently than boys.

Why?

Is there something about adolescence—high school, hormones, homework—that kidnaps a mind in bloom, transforming it, sometimes out of the blue, into a mind consumed by madness or gloom?

There's little doubt that genes are partly to blame—both schizophrenia and depression run in families. But studies have shown that only 50 percent of the risk of developing either illness can be attributed to genetic legacies.

There's also no doubt that stress plays a part. All teenagers cope with the high drama of ninth grade, but some evidence suggests that adolescents overall are more reactive to stress. A human teenager's blood pressure rises higher in stressful lab situations than that of his parents or younger siblings; teenage rats under stress startle more easily, dig more nervously, lose more weight, and cower in corners longer than their younger or older counterparts. If you have identical twins, both with the same genetic leanings toward depression or schizophrenia, many scientists believe it could be teenage stress—or perhaps a particular teenager's outsized perception and reaction to stress—that helps tip the balance, leaving one twin sipping mocha lattes with friends at Starbucks, the other sick and alone, caught in the grip of grandiose delusion or mind-numbing depression.

INCREASINGLY, the teenage brain itself is seen as a culprit. So much of the brain is still changing in adolescence—in particular the prefrontal cortex, which has been implicated in both schizophrenia and depression—that clues to both these devastating and elusive illnesses, many now believe, may lurk amid the transformations in a teenager's brain.

Francine Benes at McLean Hospital, for instance, who has studied schizophrenia for years, thinks myelination may play a role.

As adolescence progresses, nerve cells in key brain areas get a layer of myelin, a coating of fatty insulation that both speeds and smoothes signals between neurons. Usually, such improved communication is a good thing, and may contribute to leaps in cognition during teenage years. But if that improved communication connects areas with abnormalities, brain function can get worse, somewhat like running a faster train over a defective track.

Specifically, Benes thinks that the streamlined connections may unmask a dysregulation with part of the brain's inhibitory GABA system, leaving schizophrenics unable to calm their brains and filter out the bombardment of confusing signals from the outside world.

Dopamine, too, may play a part here. During adolescence, nerve fibers from brain cells that release dopamine proliferate in parts of the frontal cortex, and elsewhere, and Benes's lab has shown, they increase their connections with GABA neurons. Dopamine works to dampen the inhibitory GABA cells, so with more dopamine you get less overall inhibition of signals in the brain. Schizophrenics, who may have irregularities in this system, Benes says, could then be left with an "unfettered flow of information from the outside world inward—overwhelmed by sensory stimuli."

Add to that the trigger of stress, known to increase levels of dopamine, and the situation is exacerbated. (Most antipsychotic medications block dopamine, so such symptoms, according to this model of the disease, would be lessened.)

Benes believes there are a number of possible, but related, routes to the development of schizophrenia, all connected to the changes taking place in the brain, as well as increased stress during adolescence. "There are multiple developmental changes taking place in adolescence that provide a window of opportunity for the vulnerability to schizophrenia to present [itself]" says Benes.

Other researchers have found evidence that implicates another natural process that we now know is taking place in the teenage brain: pruning. As mentioned earlier, over the course of adolescence, the average teenager loses about 15 percent of his excess cortical gray matter—the branches and cell bodies of the neurons.

But those who develop schizophrenia, the researchers have found, lose as much as 25 percent. Could this incapacitating illness be brought on when, because of genetic or environmental influences, this normal pruning process goes haywire? Some of those doing the research, including Judy Rapoport at the National Institutes of Health and Paul Thompson at UCLA, see possibilities.

"When so much is changing, when there are so many moving parts, something can go wrong, something can break," says Rapoport. "We see this massive loss of brain tissue, but then you have to ask: Is that a cause or a symptom? Whatever signal there is that makes synaptic pruning happen normally may malfunction. Or it could also be protective. It could be the brain's way of getting rid of things that are wired incorrectly."

David Lewis, a neuroscientist at the University of Pittsburgh, investigates the intricacies of schizophrenia at the molecular level. After recently reviewing much of the literature on its causes, he believes, as many do, that the disease involves the prefrontal cortex, a key brain area still being developed in adolescence.

Structure irregularities have been found in the prefrontal cortex in schizophrenics, and those with the disease have symptoms that in many ways mimic those with prefrontal cortex damage. Many are unable to call upon their working memories, to put events in context, to make sense of the incoming reality, and hold a thought—all functions linked specifically to the dorsolateral prefrontal cortex. Most schizophrenics readily admit they are overwhelmed and confused by an onslaught of unconnected information from the outside world.

And we now know that the prefrontal cortex gets substantially refined late in adolescence, a time when the bulk of psychosis emerges.

For now, David Lewis says, much of the thinking about schizophrenia is largely divided between two possibilities: Something happens in the womb or at birth—a mother's case of the flu, incorrect brain cell migration, a difficult delivery that deprives the brain of oxygen—that causes a defect in an area such as the prefrontal cortex, which becomes apparent only when that brain region fully matures in adolescence. Or, one of the natural developmental processes taking place in the adolescent brain, such as pruning, itself goes awry.

"Personally, I'm agnostic about which one it might be," says Lewis, "and they are not mutually exclusive. But I am certain that understanding how the brain changes during adolescence will give us clues."

MARIA was in college when she heard her first strange voice. As she walked through the campus of her university in Australia, she saw two of her art professors turn a corner and approach her. As they got closer, Maria heard one urgently whisper to the other one word: "Ears!"

To Maria, it was a signal. It meant the professors thought she could hear what they were saying and they were talking about her. It meant one was warning the other to be quiet because "ears," which Maria thought was a reference to herself, was approaching and they had to keep what they were saying a secret.

Looking back now, Maria knows it wasn't true. The professors were there but neither whispered "Ears!" And it's highly unlikely they were talking about her. Maria now realizes that on that bright morning she was falling into the grip of devastating psychosis, an illness that would leave her unable to discern real from nonreal, there from not there, voices from silence.

"I got very, very sick," she said when I spoke with her not long ago. "But at that point, I didn't know I was sick."

As far as Maria can figure it, the first stirrings of the severe psychosis that was to come began when she was sixteen years old. At that point, she recalls, she started "getting confused," getting "mixed up thoughts," getting angry at those mixed up thoughts.

By age nineteen, the illness was closing in. She was living with a young man—the relationship was somewhat abusive and quite stressful—and as months went by, she slid deeper and deeper into her own world. A painter, she spent hours by herself, but when friends stopped by, she was increasingly consumed with confused and paranoid thoughts. "I began to think they were talking about me, looking at me funny; that they thought I was strange," she says. "At times, I wouldn't hear what they were saying; instead I would focus on *how* they were saying it, what they looked like, the process. Then I would get lost and confused. I didn't know what was going on."

Once, after a bad spell, her mother took her to a psychiatrist who dismissed their concerns, telling Maria she simply had "a lively imagination" and attention deficit disorder. He prescribed amphetamines, a standard treatment for ADD. To quiet her mind, she began to smoke a lot of marijuana, a drug, she says, that brought a "more distorted reality."

By the time she was twenty, Maria was quite ill. She couldn't sleep and often felt cut off from reality. Not only did she believe professors were whispering about her, but she thought they were signaling her with special messages on classroom blackboards. She heard disturbing noises and a stream of commentary about her work from people, who, it turned out, weren't there.

"When I started doing something wrong, there would be this noise, this bashing of drawers. Then, if I did something right, I'd hear 'Now, that's better,' " she says. "I thought people could hear my inner thoughts, which seemed to be floating all over; other people's thoughts were floating in the air, too. I would look at other people and not understand how they could concentrate with all those thoughts in the air."

One afternoon, she broke. Walking down a hallway at her university, she became convinced she was God; then, that she was Jesus. She told herself, "No, he died on a cross, so you will die, too." She panicked, convinced she would die any minute. By the end of the day, Maria was in a hospital, where she stayed for four months. Her diagnosis: schizophrenia.

And that was just the start. For years she suffered from on-and-off psychosis and medications that didn't work that well. Eventually, though, Maria got what she considers her first lucky break. She was put in a special new program in Melbourne set up to treat young schizophrenics.

The program takes an unusual and radical approach to schizophrenia. Run by researchers at the University of Melbourne, it tries to catch and treat schizophrenia early with low doses of antipsychotic medications to see if the disease can be delayed or waylaid before severe symptoms emerge.

For many years schizophrenia was considered a degenerative disease, much like Alzheimer's. But because there's no progressive tissue damage, and because a percentage of schizophrenics improve somewhat as they age, the disease is no longer seen as following an inexorable downward spiral. Some scientists now believe that if caught earlier and properly treated, the ravages of schizophrenia may be lessened.

The Australian study, begun in 1996, has already produced provocative results. According to Patrick McGorry, the psychiatrist at the University of Melbourne who heads the study, only three out of thirty-one teenagers who were treated early with antipsychotic medications along with tailored psychotherapy went on to develop full-tilt psychosis. In a control group that received only therapy, ten out of twenty-eight became fully schizophrenic.

Not all agree with this approach. Although a similar study is now going on at Yale, the idea of treating teenagers with powerful drugs before they have definitive symptoms has been controversial in the United States, where McGorry said "it's perceived as a kind of conspiracy of the drug companies."

But McGorry takes a different view. He believes that by setting up special health centers where teenagers can come "without stigma," as they've done in Australia, and getting teenagers considered at high risk for developing schizophrenia proper treatment early—perhaps around the time when Maria first felt her confused thoughts closing in—we may be able to dilute the virulence of the disease. McGorry even says that therapy alone, if extensive enough, might also work eventually. But he and a number of others are beginning to wonder now if the antipsychotic medications might help because they are somehow "neuroprotective"—that is, they actually work to limit damage to brain cells.

McGorry suspects that much of schizophrenia develops when

some natural developmental process in the teenage brain get derailed, either because of genes or the environment (which might include heavy drug use in his view).

"If you look at the development of the teenage brain, there is a huge restructuring during this period. There's the normal process of cell death and removal. That's under genetic control, but it's also influenced by a lot of internal and external factors," he says. "We know these things are going on. But in schizophrenia, something in this normal adolescent development goes awry.

"Most of these kids have normal childhoods," he added. "That's why parents are so devastated. One day they have a normal kid, often a high-achieving kid, and then something happens. Something quite profound is happening in adolescence."

The difficulty, of course, is figuring out early on which kids are headed into the depths of delusion and which are just having an extended bout of garden-variety teenage weirdness. McGorry says there are often signals that can be detected if you're looking for them, including increasing social withdrawal, sudden poor performance in school, whispered voices, or suspicious feelings. He says his group has been able to accurately predict which teenagers are headed for full psychosis in 80 percent of cases that are referred to their clinic for evaluation.

With this approach, a young teenager such as Maria might have been diagnosed and treated as soon as her confused thinking became persistent, before her illness progressed to the point where reality had slipped away, long before she thought she was God.

"This is really the opposite of how schizophrenia has been treated in the past," said McGorry. "We've always waited until it got very severe and then tried to treat it. But that might be too late."

Although Maria, by the time she joined the program, was too old to officially take part in the early-treatment studies, she nevertheless benefited greatly from its aggressive approach to her illness. After spending months in adult programs where, she says, they locked her up and "jammed" needles into her, she found the teen program to be more compassionate—and effective, treating the illness more matter-of-factly, as something that can happen to a brain at adolescence, but that can be managed. "It empowered me," she says.

Depression and Anxiety Rear Up

In the bowels of the National Institutes of Health in Washington, a search is under way to detect early stirrings of two other mental disorders that begin to rise at puberty: depression and anxiety.

Researcher Danny Pine is looking inside hundreds of living teenage brains to see if he can pinpoint when and how they fall prey to deep depressions or crippling anxiety, including social phobia and many classic panic attacks, which also start up at puberty—the kind where hearts race, palms sweat, and you feel like you can't breathe, or that you're going to die for no apparent reason.

Unlike schizophrenia, whose onset typically comes anytime in the long stretch of adolescence, depression and anxiety are more directly linked to the more precise biological moment we think of as puberty. Rates of major depression, for instance, rise according to how far along a child is in puberty, rather than how old that child is.

To find out why, Pine is putting adolescents and preadolescents—depressed or anxious as well as those who feel perfectly fine—into an fMRI scanner to see how their brains respond to images of happy, afraid or angry faces. He's also putting a group of willing grown-ups through the same process, to act as a control group.

Pine chose angry and fearful faces, in part, because a fair amount is known about how a normal brain processes those emotions, in a complex chain reaction that's thought to involve the amygdala and the prefrontal cortex. It's also true, he says, that depressed teenagers tend to be more angry and irritable than depressed adults, and their symptoms often show up after heated exchanges with their parents.

What Pine wants to know is this: As children go through puberty, do their brains respond to angry faces differently? Do they use different parts of their brains to process anger than grown-ups? Do teenagers showing signs of depression or anxiety use different brain parts or use the same parts differently from those kids who feel fine?

In addition to scanning their brains, Pine is measuring levels of their estrogen, androgens, and, for stress, cortisol. (Interestingly, girls are more likely to show psychiatric symptoms in response to a stressful event after puberty than before, Pine says.) He is investigating whether those ever-elusive hormones are any different in kids with problems than in kids who don't have problems—and whether different levels of hormones coincide with different activity in various teenage brains.

Certain things are known about how depressed or anxious teenagers behave. Not surprisingly, depressed teens tend to focus on the negative. Tell them they got an A on a test and they will tell you it was a fluke. Give them a list of happy words, like prize and win, and a list of neutral words such as boat, and a list of negative words, and they will nearly always remember the negative ones more easily. Show them a picture of Glenn Close being angry and Keanu Reeves being happy and later on they're much more likely to remember the picture of Glenn Close.

Anxious kids, too, follow certain patterns. It's thought that certain anxiety disorders may come about through a dysregulation, perhaps a hypersensitivity of the HPA axis, the hypothalamo-pituitary-adrenal axis. Some type of stressor—eighth-grade math, a rude remark from someone in gym, a request to take out the trash—sets off a cascading release of hormones, including a rush of cortisol into the bloodstream, that triggers the body's fight-or-flight response, sweaty palms and all.

And it's already been shown, Pine says, that teenagers who are anxious tend to be hyperalert to signs of danger. If you show them a movie of a big crowd in which one face is obviously angry, they're considerably quicker at finding that angry face than kids who are less troubled by anxiety.

Still, we have to pay homage to the slippery dance of nature and nurture. Susan Nolen-Hoeksema at the University of Michigan, who has devoted her career to trying to figure out why young teenage girls get depressed earlier and more often than young boys, says her most recent research finds that the pervasive idea that all teenage girls are more prone to depression may be a "cultural myth" and that a small portion of severely depressed girls may account for the overall increase.

But even this thought leaves open the question: What is going on in the lives and brains of that segment of girls as they hit age thirteen that sends them into a downward plunge?

Do girls get more depressed because they're at the mercy of wildly fluctuating estrogens? Do girls get more depressed because their brains are set up differently to begin with? Do they get depressed because they're repressed in math class or at home? Or do girls get more depressed because, as some anthropologists who study monkeys maintain, they need help from trusted friends to raise the kids later, so they've evolved to be more group-oriented, more invested in social bonds, and, therefore, more devastated when those bonds are broken?

Nolen-Hoeksema believes girls have an increased chance for a build up of "negative events." Sexual abuse can lead to depression, and sexual abuse is more common in girls than boys. Getting stuck without options in the world can lead to depression, and that, too, is more common in girls. Those girls who are "early maturers," the ones who get their periods and their breasts in elementary school, for instance, are also at high risk. The world treats them differently because they look older, parents treat them differently, boys prey on them and they usually don't have other comparable girls around whom they can lean on, trust, and get support from.

There are many other cultural and environmental reasons behind depression and, undoubtedly, there's a genetic component as well. It may be that some teenagers have a biological vulnerability not to depression per se, Nolen-Hoeksema says, but to difficulties in regulating mood, in calming down, in males and females. Faced with that biological leaning, boys and girls may respond differently when stressed. Males, when upset, tend to distract themselves, perhaps go out for a drink. Females, who are more often stuck in their environments or feel they are, tend to "brood and ruminate more." The result? More boys take what Nolen-Hoeksema says is the "more socially accepted" road of becoming alcoholics, and more girls get depressed, the culmination of an ever-mutating mix of biology and culture.

But, she's quick to add, "even that doesn't explain all of this."

Danny Pine, for his part, believes that to explain all of this we have to watch the adolescent brain at work, from the inside out.

After seeing thousands of kids who were severely depressed or suffering from debilitating anxiety, he has turned to brain-scanning because, he says, even after all these years, "we have only a rudimentary understanding of what causes psychiatric disorders." By and large, he says, traditional methods, the "huge effort to use clinical practices and research to answer those questions have consistently come up short."

But brain-scanning, so far, has left us with a confusing picture. Some studies have found that certain brain parts are bigger in teenagers and children with depression or anxiety, while others have found those same parts are smaller. As a result, Pine has embarked on his large-scale, years-long look at the inner neural workings of adolescence, looking not just at what parts are bigger or smaller, but how those parts interact with each other in living, growing teenagers.

To him, there is an urgency to all this.

Although he says it's hard to claim, as some do, that psychopathology among teenagers overall is up—there's less of a stigma, so more might be coming for treatment—there is one statistic that has Pine worried. While it's dipped slightly recently, over the last 50 years the rate of suicide among teenagers has steadily gone up. "In that way, you could say things have gotten worse," he said.

Holding On to Normal

As Pine and the dozens of other researchers who are poking around in the teenage brain try to figure out what makes that brain go wrong—or right—teenagers themselves continue, in their own various ways, to try to figure it all out, too.

For those like Maria, who has battled debilitating psychosis for more than ten years, the figuring out is still tough. She's living on her own now, and she works part-time. She's learned to recognize when her mind is slipping into confusion and tries to fight it off. "When those strange thoughts start to come, I say to myself, 'hold on, it's not real; that's just your mind trying to bullshit you again,'" she says. But none of that is easy; her life, while considerably better than it was, remains a moment-by-moment risk. "I just take it day to day; that's the only way I can make it through," she says.

For teenagers not battling such severe mental illness, the daily struggle is not so urgent. But no one ever said adolescence, in whatever form it takes, was easy. Even among those blessed with brains functioning in normal ranges, there's the very complicated problem of growing up.

Stuart, a perfectly normal teenager from Maryland, says he went through a time when he had "unpredictable emotions," when he would get into arguments with his parents over very little. He sees some friends get swallowed whole by adolescence, kids who "do a lot of drinking and drugs."

"Sometimes it seems like the goal of adolescence is to get out of it, an urgent need to grow up," he says. "A lot of kids I know struggle with that. Sometimes when you're a teenager, it's hard to feel comfortable with who you are at the moment and I think that causes a lot of the problems."

But at age eighteen, on the cusp on manhood, life for Stuart is sweet. The high-pitched emotions and inexplicable wrangling with his parents—and himself—have subsided. He has found his passions, basketball and the viola. He's comfortable in his skin, proud of his newfound stability and sense, and excited by the possibilities in front of him. He's also particularly happy to report that his brain has somehow, as he puts it, "stretched." His thoughts are sharper, deeper, and richer. And that feels good.

"I think one of the biggest changes I notice is that my mind seems to see things in a more complex, complicated kind of way now," he says. "It's like for the first time, my brain can ask 'what if.' "

Chapter 14

Coming of Age

On the Path to Maturity

A TEENAGER'S brain is still unfolding. Like its owner, it seeks its way, reaching here, stumbling there, pushing out, and pulling back. And within such flux lurks promise. Just as the Chinese symbol for change means both peril and possibility, the teenage brain, too, holds both.

"When I hear that the teenage brain is still changing so much, I get worried," says Barbara Pedley, a neurologist and a mother of two teenagers. "It means that more things can go wrong." "What does this all mean?" says neuroscientist Jay Giedd. "In the end, it means that we shouldn't give up on any teenager, there is hope." As we've seen, the remodeling of the adolescent brain—a brain that science had considered largely finished—spreads over such a wide range of systems that we need to rethink how we think of teenagers altogether. Over a span of roughly ten to twelve years, the adolescent brain, through a series of sometimes subtle and sometimes breathtakingly dramatic shifts, is transformed from child to adult. The gray matter of an adolescent's frontal lobes grows denser and then abruptly scales back, molding a leaner thinking machine. The teenage brain fine-tunes its most human part, the prefrontal cortex, the place that helps us cast a wary eye, link cause to effect, decide "maybe not"—the part, in fact, that acts grown-up.

The brain of a teenager undergoes a proliferation of connections for dopamine, a neurotransmitter important for movement, alertness, pleasure—high levels that may have evolved to help adolescents of many species take the necessary risks for survival, from exploring new fields for food to asking that saucy young girl to dance. The long, thin arms that connect brain cell to brain cell are coated with insulation that speeds signals in brain regions devoted to such fundamental capacities as emotions and language. The cerebellum, associated with understanding social cues and even jokes, blossoms and then consolidates, a process that spans adolescence and continues well into our twenties. Brain chemicals that help determine sleep patterns shift in adolescence—perhaps a legacy from a time when the quick-eyed young needed to stay up into the night to protect the rest of us.

Recognizing the Change

Such cerebral transformations are crucial to the development of a normal, average teenager. And recognizing and understanding these changes is a crucial—and until now largely missing—step toward helping parents and kids deal with the day-to-day doings of the normal, average teenage life.

The research I did for this book surprised me—and changed the way I think about my own teenagers. The changes occurring in the teenage brain are so profound and wide ranging that they poke into nearly every area of behavior, adding a new and abundantly welcome perspective to the enduring puzzle of adolescence.

Why is that teenager still curled in a motionless ball, sound asleep at one P.M. on a Saturday? Doesn't he know there are clothes to be picked up, homework to finish, papers to compose? Is he lazy? Is she dead? Do they hate us?

Well, maybe not.

Why is that teenager sneaking in at three A.M. with the breath of a sailor and the glare of disdain? Is it because, as one mother of two teenage boys told me, it was so easy to fear at times like this, you're simply "raising little felons?"

Well, maybe not.

The new brain science tells us that many of the perplexing, aggravating, and, at times, delightfully surprising activities of a teenager are just that—the perplexing, aggravating, and delightfully surprising activities of a normal, average teenager. And knowing that an upheaval occurs in the teenage brain—naturally and inevitably in your kid and the kid down the block—can help. Indeed, many of the neuroscientists who have children of their own have embraced this knowledge—and found ways to use it in their own lives.

Adjusting Expectations—and Actions

Deborah Yurgelun-Todd, the neuroscientist at McLean Hospital, says any new evidence that bolsters the idea that adolescent behavior is a necessary, built in—and temporary—phenomenon is a parental blessing. And it means we ought to take the long view. No one is saying that parents should back off. If a teenager is having real emotional difficulties, it's imperative to step in and get help. This doesn't mean, either, that teenagers are incapable. They're smart and able—and still need their size-10 feet held to the fire. The fact that their brains are still developing is not, as one father familiar with the new research told me, "an excuse not to take out the trash."

But, with evidence suggesting that adolescents sometimes think differently than we do, perhaps we should consider a few adjustments in our own actions and expectations. Yurgelun-Todd's research found that young teenagers often get confused when trying to decipher the emotions of others. Lacking full frontal-lobe function they may, more often than adults, react with parts of their brains primed for fear and alarm. And if teenagers are befuddled and on edge—by design—we should take that message to heart. In Yurgelun-Todd's own house—to slice through the fog and get her kids to actually hear what she's saying—she's trying new tactics. Instead of asking her daughter to brush her hair, pick up her room, and empty the dishwasher all at once—and being met with the blank stare of inaction—she now asks for only one thing at a time, slowly, calmly, ready to repeat as needed, a tactic other parents of teenagers told me they also try, without knowing why. It's a small step, but it's surprising how well it works.

Providing a Dash of Foresight

Peter Jensen, the child psychiatrist at Columbia University and father of five himself, acknowledges that the new discoveries about the continued change in the teenage brain, as well as evidence suggesting we learn best when exploring varied territory on our own, have changed his way of thinking, too. The still-developing state of an adolescent's prefrontal cortex, Jensen believes, means teenagers don't always see the consequences of their actions and parents sometimes do need to step in—like a surrogate frontal cortex—to provide a dollop of foresight. And the best way to accomplish this is not necessarily with a jackhammer. With teenager number five, Jensen has learned to sit back a bit, give subtle nudges, hint at possible outcomes perhaps, and then let his daughter take her own newly connecting frontal lobes for a spin.

"When my oldest child was a teenager, I was always trying to be in control, always trying to be her forebrain," he says. "I am trying to raise the fifth one differently—giving structure, but also giving more choices to her own forebrain, choices she can make herself."

They're Not Adults

In dealing with teenagers, we need to focus not only on their sometimes odd perceptions of the world, but on our own, as well. After watching an adolescent grow a foot in a year, it's difficult not to think of that towering teenager as a complete and polished grown-up.

"With teenagers, it's especially hard to remember that their brains are developing because they look like adults. I have to crane my neck to yell at my son," says Chuck Nelson, the developmental neuroscientist at the University of Minnesota. "But even though teenagers have the bodies of adults, they are not adults. We must keep that thought in mind—if we can."

Indeed, the new evidence that the adolescent brain is still a work in progress should make us all a bit more forgiving of teenagers and ourselves.

"I think this new knowledge will help parents be able to say 'Hey, maybe my kid is not as crazy as I thought. Maybe it's natural and I can wait this out just like I waited it out when the kid was younger and having temper tantrums,'" Nelson says. "Maybe it will help parents realize that it's just a stage, too. The areas of their brains will eventually mature to the point where they'll be more adult. Maybe, if we think of that, we can be more charitable."

As a father of a teenager with a still-maturing brain, of course, Nelson concedes that such charity is not always easy.

"You know how it goes," he said with a laugh. "I have a split in my own brain. When my son is acting up, one brain hemisphere is understanding; I know what is going on in his brain can explain his behavior. But the other hemisphere wants to say to him, 'Hey, you're being a big jerk.' Just the other day, it dawned on me that my son and I have had a lot of conflict lately—yelling all the time—and it was the same thing I had done with my own father at his age. It made me realize that it was probably natural and, thinking that, for a whole eight hours, I was real understanding, real nice.

"But," he said, "who knows what will happen today? Sometimes I treat my son as he should be treated, based on what I know about brain development, and then, you know, he pushes my buttons."

PLANNING FOR RISKY BEHAVIOR

One way teenagers push our buttons, of course, is acting downright brainless, taking what seem to us to be dumb and unnecessary risks. But this, too, can be looked at in a different way. If, for instance—as everyone from human brain scientists to those who deal with adolescent rats and monkeys tells us—teenagers are not only attracted to risky behavior, but such behavior is a natural and necessary evolution in their development, it's time we understand that—and expect it.

This doesn't mean we should purposely let hapless teenagers be foolhardy and reckless. But if teenagers, in varying degrees, have been acting this way across cultures and species since the dawn of time, why is this such a hard fact to swallow? If we expect such behavior, we can plan for it. A few states, hoping to curtail risky driving for instance, have decided to limit the number of teenagers that can be in a car at the same time; others have lengthened the time a young driver has to have a learner's permit and drive with an adult before getting a license, a step that has already reduced the number of accidents and deaths. And if teenagers who sexually mature earlier than their peers are more at risk of getting into trouble—their strong new urges perhaps not yet curbed by frontal-cortex caution—maybe we should plan for that, too. We could easily focus more attention and help on those kids—rather than just being mystified, angered, and horrified by their behavior.

A Chance to Sleep

Or, on a more mundane level, if teenagers sleep until noon in part because shifts in brain chemicals make them drowsy later—and because they're deeply sleep-deprived from getting up at dawn to get to high schools that begin ridiculously early—why can't schools start later? And shouldn't schools at least think about the possible impact on sleep before assigning piles of homework every night, or not-so-subtly insisting that all kids—for those all-important college resumes—be involved in endless lists of organized extra-curricular activities?

Teenagers are fired by surges of natural, exuberant energy, but they also still need more sleep than adults. Teenagers should be exposed to as many challenges and new experiences as possible, but do colleges think that when they get an application from a kid who's been class president for three years, editor of the yearbook, captain of the soccer, fencing and swim teams, a prize-winning oboe player and playwright, as well as CEO of his or her own small computer graphics company that they are getting an application from anyone other than a completely exhausted human being?

Taking some steps, a handful of schools have tried starting later, and it works. Kids in those schools are not only more awake in class, but less churlish. After reviewing findings from adolescent sleep research, even the U.S. Navy decided last year to let new recruits—most of them still teenagers themselves—stay up later, sleep later, and get eight hours of sleep instead of six, a model that used to be standard but had been forgotten. If the military can do this, why can't the rest of us?

Turning Down the Pressure-Cooker

In response to our new knowledge about how the adolescent brain grows and develops, it may also be time to give teenagers a wider definition of what success at this age means, give them more wiggle room to make their own mistakes and come up with their own answers. Why not lighten up on the oversupervised, overstructured activities and get-into-a-good-college-or-else-you're-dead culture we've created and, instead, figure out ways to let teenagers have some space to find their own path? Maybe we should stop robbing them of the time they need to take risks and roam intellectually, physically, and emotionally. In our supercharged world, have we left them with only the time and space for the quick-fix thrills of sex at twelve or a handful of Ecstasy?

Suggestions have been made: more varied and available internships, work experience and job training, or travel in the real world—and time built into overregimented academic careers to do those things, and still get into college. In Europe, it's accepted and far more common for teenagers to take a year off between high school and college—a so-called gap year—when they can take time to look at the world and themselves. Some competitive schools in the United States have been brave enough to actually stop adding advance placement classes, which many educators complain can sometimes be little more than survey classes requiring mountains of memorization, thus freeing up more time in class to dig deeper, allowing students to think and ask questions that won't be on the test.

Other schools, faced with frightening levels of teenage drinking, have—in what seems a useful stretching exercise for developing frontal lobes—turned to teenagers themselves to come up with solutions. And many of them—requiring parents to drop off kids at dances, establishing more safe houses where parents agree to be home when parties are taking place—are astonishingly smart and certainly worth a try.

I would heartily vote for all that. But I would also vote for parents like myself to take a deep, collective breath. We need a new mindset. Armed with our knowledge about how the adolescent brain continues to develop, perhaps we can cut teenagers and ourselves some slack. If nothing else, the old instinctual knowledge familiar to our grandmothers "They'll grow out of it"—now has a modern, scientific foundation and deserves a second look.

Maturity—and Biology

Teenagers, even the very best of them, can be crazy. But confronted with that craziness, the new brain science gives us another arrow in our quiver of response. Faced with that occasional insolent stare (or worse) in my own house, I can now think what a difference that time will make—time for actual brain development, time for the brain to grow, to change, to mature, time for some spit and polish on those prefrontal cortices, time for teenagers to become who they really are.

Indeed, many of those who think about adolescents in the aggregate predict that the new findings about the teenage brain, as they sink in, will change our most fundamental notions of maturity.

Robert Blum, the pediatrician and professor at the University of Minnesota who analyzes data from the nation's largest ongoing survey of adolescents, says that if science continues to indicate that teenage brains are less complete than we thought, it's likely to have "huge implications" across the political, social, and legal spectrum, from the question of whether to allow a young teenage girl to get an abortion without her parents' knowledge, to whether to prosecute a young teenage boy as an adult.

"This will change the whole debate about adolescence in directions that we can only begin to guess," Blum said. "We've gotten into the idea that it is mostly the environment that has an impact, and we have swung very strongly away from any biological explanation for things. But I do think this new research will right the balance—biology and genetics have to be part of the equation. And it will increasingly affect our understanding of teenagers and influence policy in very a complex way. It raises the question: What does it mean to be mature?"

Elizabeth Cauffman, the longtime researcher of adolescent behavior at the University of Pittsburgh, says she fervently hopes the new findings by neuroscientists, combined with similar research by child psychologists and sociologists, will influence recent trends in juvenile justice, where the country is still debating how to deal with young teenage offenders, many of whom are being tried as adults.

"Of course, it depends on what the government is willing to hear," she says. "But I would hope that, in the end, the law would be more sensitive to these developmental issues."

THOSE who counsel teenagers fully expect that new and continuing discoveries about the brain will alter the landscape of their own field, as well.

David Fassler, chairman of the council on Children, Adolescents and Their Families for the American Psychiatric Association, who treats teenagers in his practice in Vermont, says the new findings have already changed his perspective.

As science continues to show how behavior and brain structure dance in tandem—anatomy influences emotions and experiences, and emotions and experiences, in turn, alter the fundamental architecture of the brain—we do have to be even more concerned, he says, about certain kinds of experiences teenagers may have.

"I worry even more about those kids who have repeated negative experience, who have been abused or bullied all through high school—actions that have been associated with a neurochemical state—because these experiences can increase the risk of subsequent similar events," Fassler says. "Once the brain develops pathways for responding to external stimuli, it's easier to re-create that state when that individual experiences similar stimuli in the future. That means adolescents who were bullied, from a neurophysiological standpoint, may be more likely to respond in a similar manner when they are adults."

If an adolescent's brain is still forming such long-lasting pathways, then Fassler also worries about enduring effects of alcohol, drugs, and even prescribed medications. Many new medicines enable teenagers to have a more normal passage to adulthood—those with attention deficit disorder, for instance, who are left untreated are more likely to have problems with school, peers, and substance abuse. But the long-term impact of powerful psychotropic medications that are increasingly used on teenagers remains unknown. "If there are brains changing, we have to be as careful as possible with the medications we use," Fassler says.

For Teenagers, Awareness Does Help

Still, if brains are not fixed in place, that increases the possibility that what we do to help may actually work. And kids themselves, Fassler says, will benefit from knowing what's happening in their own brains.

"For the teenagers, this knowledge can be empowering," he says. "It really helps kids to tell them that this problem they have, if it is depression or whatever, is not their fault. We can tell them that the problem is in how their brain is wired, and we can work together to figure a way around it, maybe with therapies or medications or with learning to recognize early warning signs. Knowing how all this works on a neurophysical level, and knowing that there is potential for change, can give kids a sense of mastery.

"And I think it's also helpful for parents to know that teenagers are still in the process of brain development," he added. "It increases the opportunities to help teenagers modify certain emotional or behavior patterns before they actually enter adulthood. Particularly if there are disruptive or dangerous behavior patterns, it gives parents a sense of hope that some of these problems may not be permanent."

Fassler, however, remains concerned about the environment that teenagers and their families find themselves in today. Like many who work extensively with adolescents, he is convinced that teenagers today are under more stress than previous teenagers—and many have less stable home and community lives to offset it. And those kids without nurturing childhoods enter the potentially turbulent territory of adolescence with what he calls a kind of "pseudo maturity," leaving them more "malleable and susceptible" to the tempting tugs and pulls of adolescence.

"We need to give kids exposure to a range of stimuli," he added. "But we can't give them stuff too early that's way over their heads, because they will lack the context and they will be unable to learn the lesson. But we do need to challenge them at, or slightly above, their capacity in a variety of spheres."

Low Hurdles, High Hurdles

Such thoughts, of course, have hardly escaped those who work with teenagers in schools day in and day out, like Ken Mitchell, a longtime teacher, who is now principal of Bell Middle School in Chappaqua, a suburb north of New York City.

"I think you really have to make a distinction between early adolescents and older ones," Mitchell says. "It's just like you have high track hurdles for high school but you would not use them for middle school. They have their own hurdles but they are smaller. The hurdles do have to be appropriate."

To Mitchell, increased knowledge of how the brain works should, he hopes, also lessen growing anxieties over grades, college, and the burgeoning number of standardized tests.

"It's really important for the general public and parents and teachers to know, in particular, that in early adolescence the frontal lobe is not finished," he says. "It's important to think that if you have an adolescent in a seventh-grade science class and he or she is having difficulty with abstract concepts, it may have nothing to do with intelligence, but may have to do with brain development and developmental readiness. And if we can get that across, maybe we can diffuse some of the anxiety out there.

"I think the brain learns, in part, by seeking relevance—a

word that has been overused—but I really think that young adolescents have so much energy and they are trying so hard to figure out the world," Mitchell added. "They are trying to find patterns and get the information they need to survive. That means we have to give them real consequences. And when we give them information and they ask us why it's important, we need to be able to answer with something other than 'because you'll need it in ninth grade.' "

Diane Ravitch, an educational historian and an assistant secretary of education for the previous Bush administration, says she, too, believes it will be highly beneficial if the new science about the teenage brain erases the misguided but popular thought that all important brain development occurs between ages zero and three.

"If it means that we don't give up on kids, that's a good message, that's a hopeful message," she said. Still, she worries that as such science trickles into classrooms, it may be misinterpreted, misapplied, and end up another fad in education, which "is already a graveyard of fads."

"We all want to find the feather that will make kids fly and it can be terribly upsetting to find out there is no one magic feather," she says.

It would also be an enormous mistake if the new science is used as an excuse to forget what we already know about how teenagers grow best—with parents who care and provide models of behavior, as well as more traditional methods of brain enhancement, such as practice, practice, and more practice.

Howard Gardner, the Harvard professor of education best known for his theory of multiple intelligences, says that when any new science emerges, it can, when misinterpreted, narrow rather than broaden our understanding of what makes humans do what they do, a still wildly complex conundrum at best. While he considers himself an "unabashed enthusiast" for the new efforts to poke into the teenage brain, he is concerned that if we focus on its discoveries in isolation, we'll forget that adolescence, like all of life's stages, remains a mysterious beast, a "fuzzy concept."

IN FACT, those working on the edges of neuroscience today, perhaps more than anyone, continue to embrace that fuzziness, with each new finding only deepening their appreciation for the enduring puzzle of the human mind, adolescent or otherwise.

I was most struck by this on the morning I spent at New York City's Bellevue Hospital watching what neuroscientists there rather unceremoniously call a "brain cutting," the formal dissection of brains to determine cause of death.

The setting could not have been more clinical. In the brightly lit neuropathology lab, Dr. Douglas Miller, lab director, stood behind a long table with a row of eight silver paint cans on top. Each can held a brain.

One by one, Dr. Miller delicately lifted the brains out of their containers and placed them on the table under the glaring lights. The brains were tan from the formaldehyde in which they had been soaking, with purple lines and folds. With his sixteen-inch serrated blade—it looked like a long bread knife—he sliced each brain in half like a cantaloupe. He then meticulously cut each brain half into thinner slices, leaving pieces spread out like slabs of ham. As he worked, Miller spoke out loud about what he found. (Most of the brains were from elderly people who had died with a variety of presumed ills—heart attack, stroke, dementia.)

Dr. Miller's task was to see if what he found on the inside of each brain matched the suggested cause of death, or whether there was evidence of some other anomaly. With each brain, he picked out tiny segments to examine more closely. There was the tiny sea-horse shaped hippocampus, repository of new memory. In the brain of one elderly man who had suffered from dementia, the hippocampus was the consistency of a small sponge, so soft and full of holes brought on by cell death that Miller could nearly poke his finger through. Next there was the amygdala, the almond-shaped seat of fight-or-flight reactions, found in one study to be on overdrive in the average adolescent. And there, laid out on the table, was a frontal lobe, regal and silent, giving no hint that, as it grows, it transforms teenager into adult.

It was—all of it—astonishing, even awe-inspiring, to see. But to me, one of the most astonishing moments came just before each brain was cut open. There was a moment—just as Miller's knife hovered above each brain—when there was a hush of anticipation, a collective holding of breath from all in the room, even the most toughened of neuropathologists. Each brain, after all, looked just like the rest from the outside. For that short moment, before it was sliced open, each brain held its secrets tight. What would they find in there?

Later, sitting in his office, Miller himself shook his head at the lasting wonder of it all. As the father of three boys, two of them teenagers, he, too, has spent a fair amount of time thinking about their brains and how they work, or don't.

"You know, from the outside, you can't tell the difference between a ten-year-old and a sixteen-year-old brain," he said. "But maybe they are wired differently on the inside. You can't see it, but perhaps there is something different in there. My own kids are good, but, you know, sometimes . . . well. My wife just called and said that some of them had just returned from their senior trip; they went to Florida—and some of them, not my kid, thank goodness, got it into their heads to trash the hotel rooms. They got drunk and they trashed the rooms. Now why would they do that? They know better. They have good brains. But they act like that anyway. I deal with brains every day and I don't understand it; it's wacky."

Such wackiness, indeed, is what continues to intrigue many of those who delve deep into the muddle and magic of the human brain.

Oliver Sacks, the neurologist and author, says it is the very incongruities and inconsistencies of the adolescent brain that pique his interest, the good mixed with bad, the up tinged with down, the promise and the peril.

"Adolescence is a time of extreme flux, and with it comes the wonder and the danger of extreme flux," Sacks said when we talked about teenagers and their brains. "This is a time when the meanings and the categories are being reshaped; you are moving from one identity to another. I can only imagine the neurological and hormonal change that comes with that flux.

"You know Goethe in later life would describe his attacks of creativity as just like falling in love as an adolescent. That ecstasy of falling in love, is, I think, a bit what adolescence is like overall. It's an erotically charged time, a passionate time, a playful time—you are trying out different styles—before you become more rigid, petrified. I would expect there would be a lot of brain shifting around at that time."

And because adolescent turbulence occurs with such regularity, Sacks says, "other cultures all recognize adolescence universally" as a distinct time of transition, a thought he believes we ought to keep in mind in our own fast-moving society.

Even the Amish and the Mennonites, who hold to traditional ways, he says, know that teenagers have an inborn need to take risks, to sow a few out-of-the-box oats. "In those communities, they often give teenagers some frontal-lobe caution to make up for the frontal lobes they don't have yet—and then they set them at liberty and encourage them to have adventures, to make love, to travel, sometimes until their mid-twenties," he says. "They assume that they will come back and be sober citizens and most do. But they seem to realize that teenagers need to tear away for a while."

Sacks, who has written extensively about Tourette's syndrome, in which people have involuntary physical tics, movements, or needs to shout out, says he also sees similarities between that disorder and adolescence. Those with Tourette's have a perfectly intact frontal cortex, but after they engage in one of their involuntary acts, he says, they often describe it as a feeling of "juvenile impulse" that they couldn't resist. Perhaps, Sacks says, this gives us a glimpse into what's happening in the teenage brain.

Sacks, along with many others, sees adolescence as one of the most necessary and crucial steps in human development—one that should not be just endured, but indulged, even celebrated.

He's still sorry, he says, that he missed much of what's regarded as traditional adolescence. Although he took risks in the chemistry lab, the rest of his teenage life, he says, was "overly inhibited."

"Looking back, I don't think that I lived out adolescence as I should have, and I think I've been making up for that now," he said. "I wish I'd been more sociable, more unbuttoned."

Of course, behind such unbuttonness lurks danger, too. As Sacks himself stresses, any period of "psychological openness" can be one of agony as well.

"I think of Kierkegaard, who used to describe himself at this time as a big question mark," Sacks said. "Adolescence can be like that. It can be good. It can be a time of passion. But it can also be a time for severe crisis, existential, neurotic, whatever you call it. It can be scary."

And in the end, inevitably, it is parents who must help teenagers navigate that unbuttoned passion, existential or not.

And it is parents, to my mind, for whom the fundamental message of the new teenage neuroscience—that an adolescent brain is unfinished and there is still time—can be most comforting.

Adolescents are still vulnerable, impressionable, raw—even in the deep inner reaches of their tangled dendrites. That means what happens to a teenager, what happens between a parent or a school or a friend and a teenager still matters—perhaps much more than we ever thought. This is, looked at one way, quite scary. But, looked at another way, it is also astonishingly good and hopeful news.

Not that long ago I ran into an old friend on a train, the mother of a teenage boy, an exuberant and engaging boy I'd known since he was a child. I had heard that he was having some difficulties—he had been caught with some drugs and had been shuttled from one school to another when he failed to settle down.

I asked the mother how he was doing now and she looked down and shook her head. "Well, I have to go up to his school again now because he's skipping classes again. It makes me crazy," she said.

But then she looked up and her face brightened. "But you know, we haven't given up on him. We're still trying everything we can. I heard the other day that scientists are now finding out that their brains are still growing and changing even while they're teenagers. Have you heard that?"

Notes

The information in this book is derived from sources that included personal interviews, research journals, scientific presentations, newspaper and wire service reports, and books.

Page references are at left.

Chapter 1 Crazy By Design

5 *Aristotle said: The Rise and Fall of the American Teenager,* Thomas Hine (Avon Books, Inc., 1999), page 37.
Shakespeare, Romeo and Juliet aside, described: The Winter's Tale, Act 3, Scene 3.

Chapter 2 The Passion Within

16 *He published a paper:* J. N. Giedd, J. Blumenthal, N. O. Jeffries, et al., "Brain development during childhood and adolescence: A longitudinal MRI study." *Nature Neuroscience*, 2, no. 10 (1999): 861–3.

18 *In one, Peter R. Huttenlocher:* P. R. Huttenlocher, "Synaptic density in human frontal cortex: Developmental changes and effects of aging." *Brain Research* 163:(1979); 195–205; P. R. Huttenlocher and A. S. Dabholkar, "Regional differences in synaptogenesis in human cerebral cortex," *Journal of Comparative Neurology* 387 (1997): 167–178.

Huttenlocher's work, in certain ways, mirrored: Pasko Rakic, Jean-Pierre Bourgeois, Maryellen Eckenhoff, Nada Zecevic, Patricia S. Goldman-Rakic. "Concurrent overproduction of synapses in diverse regions of the primate cerebral cortex," *Science* 232: (1986) 232–235; Jean-Pierre Bourgeois, Patricia S. Goldman-Rakic, Pasko Rakic, "Synaptogenesis in the prefrontal cortex of rhesus monkeys," *Cerebral Cortex* 4 (Jan./Feb. 1994) 78–96: 1047–3211.

19 *In 1987, Harry Chugani, a neurologist at:* Harry T. Chugani, Michael Phelps, John C. Mazziotta, "positron emission tomography study of human brain functional development." *Annuals of Neurology* 22 (1987): 487–97.

20 *In fact, there are some disorders:* John T. Bruer, *The Myth of the First Three Years* (The Free Press, 1999), page 85.

In one of his most recent papers, Huttenlocher: P. R. Huttenlocher and A. S. Dabholkar, "Regional differences in synaptogenesis in human cerebral cortex," *Journal of Comparative Neurology* 387 (1997): 167–78.

For that reason, he: "Connections in Brain Provide Clues to Learning," (University of Chicago Health & Hospital System, 2000).

Chapter 3 THE AGE OF IMPULSE

27 *By some estimates, the human prefrontal cortex:* Susan A. Greenfield, *The Human Brain: A Guided Tour,* (Basic Books, 1997), page 18.

28 Neurologist Oliver Sacks, the author of *The Man Who Mistook His Wife for a Hat and Other Clinical Tales* (Touchstone Books, 1998); *Neurology*: 56, no. 8 (April 24, 2001): 1118.

29 *In the 1970s and 1980s . . . Patricia Goldman-Rakic . . . and Adele Diamond:* A. Diamond and P. S. Goldman-Rakic, "Comparison of human infant and rhesus monkeys on Piaget's AB task; evidence for dependence on dorsolateral prefrontal cortex," *Experimental Brain Research* 74 (1989): 24–40.

Chapter 4 ALTERED STATES

37 *Marian Diamond:* Marian Diamond and Janet Hopson, *Magic Trees of the Mind* (Dutton, 1998).

39 *In 1964, the team published a paper saying:* M. C. Diamond, D. Krech, and M. R. Rosenzweig, "The Effects of an enriched environment on the histology of the rat cerebral cortex," *Journal of Comparative Neurology* 123 (1964): 111–20.

Greenough has shown, in various studies: William T. Greenough, James E. Black, and Chistopher Wallace, "Experience and brain development," *Child Development* 58, no. 3 (1987): 539–59.

40 *Brain measurements of professional:* Charles A. Nelson, "Neural plasticity and human development: The role of early experience in sculpting memory systems," Developmental Science 3 no. 2, (2000): 115–36.

41 *A recent study by Harry Chugani:* Harry T. Chugani, Michael E. Behen, Otto Muzik, Csaba Juhasz, Ferenc Nagy, and Diane C. Chugani, "Local brain functional activity following early deprivation: A study of postinstitutionalized Romanian orphans," Neuroimage, Academic Press, published online, March 13, 2001.

44 *Tucked at the end of one of Giedd's papers:* J. N. Giedd, J. Blumenthal, N. O. Jeffries, et al., "Brain development during childhood and adolescence: A longitudinal MRI study. *Nature Neuroscience* 2, no. 10 (1999): 861–63.

Chapter 5 Making Connections

53 *Though her sample was small, Benes published:* Francine M. Benes, "Myelination of cortical-hippocampal relays during late adolescence; anatomical correlates to the onset of schizophrenia," *Schizophrenia Bulletin*, 15 (1989):585–94.

And she found what she was looking for: Francine M. Benes, Mary Turtle, Yusuf Khan, Peter Farol, "Myelination of a key relay zone in the hippocampal formation occurs in the human brain during childhood, adolescence and adulthood." *Archives of General Psychiatry* 51 (June 1994).

54 *H.M. could not remember:* Susan A. Greenfield, *The Human Brain: A Guided Tour* (Basic Books, 1997), page 126.

56 *With split brain patients:* Thomas B. Czerner, M.D., *What Makes You Tick? The Brain in Plain English*, (John Wiley & Sons, Inc., 2001), pages 34–35.

Such split-brain patients have been known to: Principles of Neural Science, edited by Eric R. Kandel, James. H. Schwartz, Thomas M. Jessell, (McGraw-Hill Companies, Inc. 2000), page 16.

57 *Thompson's study, the basis for:* Paul M. Thompson, Jay N. Giedd, Roger P. Woods, David MacDonald, Alan C. Evans, and Arthur W. Toga, "Growth patterns in the developing brain detected by using continuum mechanical tensor maps," *Nature* 404 (2000): 190–93.

58 *Even Broca's and Wernicke's area have been:* "Principles of Neural Science," pages 13–15.

59 *Tomas Paus, a young Czech:* Tomas Paus, Alex Zijdenbos,Kieth Worsley, D. Louis Collins,Jonathan Blumenthal,Jay N. Giedd,Judith L. Rapoport, and Alan C. Evans. "Structural maturation of neural pathways in children and adolescents: In vivo study," *Science* 283 (March 19, 1999): 1908–11.

"Who's the idiot: Harry Chugani as quoted in *The Myth of the First Three Years*, ed. John T. Bruer (The Free Press, 1999), page 142.

63 *While Jay Giedd, working in:* Elizabeth R. Sowell, Paul M. Thompson, Colin J. Holmes, Terry I. Jernigan, and Arthur W. Toga, "In vivo evidence for post-adolescent brain maturation in frontal and striatal regions," *Nature Neuroscience* 2, no. 10: 859–61.

65 *In fact, scientists like David Lewis:* "Schizophrenia and Peripubertal Refinements in Prefrontal Cortical Circuitry," in *The Onset of Puberty in Perspective*, ed. David A. Lewis (Elsevier Science B.V. 2000); "Development of the prefrontal cortex during adolescence: Insights into vulnerable neural circuits in schizophrenia," *Neuropsychopharmacology* 16, no. 6 (1997): 385–95.

67 *The brain is an energy:* Susan A. Greenfield, *The Human Brain: A Guided Tour*, (Basic Books, 1997), page 27.

In fact, the experiment Patrick: Abigail A. Baird, Staci A. Gruber, Deborah A. Fein, Luis Maas, Ronald J. Steingard, Perry Renshaw, Bruce Cohen, Deborah A. Yurgelun-Todd, "Functional magnetic resonance imaging of facial affect recognition in children and adolescents," *Journal of the American Academy of Child and Adolescent Psychiatry* 38, no. 2 (1999): 3195–99.

68 *Another recent study:* Robert F. McGivern, Julie Andersen, Desiree Byrd, Kandis L. Mutter, and Judy Reilly, "Cognitive efficiency on a match to sample task decreases at the onset of puberty in children." *Brain and Cognition* 50 (2002): 73–89.

Chapter 6 THE ADOLESCENT ANIMAL

81 *Many evolutionary anthropologists:* Natalie Angier "Why Childhood Lasts and Lasts and Lasts," *New York Times*, July 2, 2002.

84 *Most developmental psychologists trace:* Stanley G. Hall, *Adolescence and Its Relation to Psychology, Anthropology, Sociology, Sex, Crime, Religion and Education* (D. Appleton and Company, 1904).

Patricia Hersch's *A Tribe Apart: A Journey into the Heart of American Adolescence*, (Ballantine Books, 1999), page 14.

David Brooks, writing in the Atlantic: David Brooks "The Organization kid," *Atlantic Monthly* 287, no. 4 (April 2001): 40–43.

Thomas Hine, in his wonderful and pointed: The Rise and Fall of the American Teenager,(Avon Books, 1999), pages 8, 16, 46–49, 51, 62, 99–104, 301.

86 *Some troubling teenage statistics:* "Child Poverty, Adolescent Birth Rate, Continue Decline," The Federal Interagency Forum on Child and Family Statistics, National Institutes of Health, July 19, 2000. According to Blum's survey, "New Study Questions Teen Risk Factors; School Woes, Peers Are Stronger Clues Than Race, Income," by Laura Sessions Stepp. *Washington Post*, Nov. 30, 2000.

Chapter 7 RISKY BUSINESS

89 *Lynn Ponton, an adolescent psychiatrist in San Francisco and the author of "The Romance of Risk": The Romance of Risk*, (Basic Books, 1997).

90 *Nearly half of all new cases:* Centers for Disease Control, Atlanta.

Some violent acts by teenagers have declined somewhat, but: Mark Anderson, "Multiple victim violence in schools rises," *Journal of the American Medical Association*, Dec. 5, 2001.

93 *Several years ago, eight young men:* M. J. Koepp, R. N. Gunn, A. D. Lawrence, V. J. Cunningham, A. Dagher, T. Jones, D. J. Brooks, C. J. Bench, P. M. Grasby, "Evidence for Striatal dopamine release during a video game," *Nature*, 393 (May 21, 1998): page 266–68.

96 *To pinpoint where that drive might:* M. T. Bardo, S. L. Bowling, P. M. Robinet, J. K. Rowlett, M. Lacy, B. A. Mattingly, "Role of dopamine D1 and D2 receptors in novelty-maintained place preference." *Experimental and Clinical Psychopharmacology* 1 (1993): 1–4, 101–109.

97 *In another experiment, Bardo:* George Rebec, C. P. Grabner, R. C. Pierce, and Michael Bardo, "Voltammetry in freely moving rats: Novelty-dependent increases in accumbal DOPAC," abstract presented at the Annual Meeting of the Society for Neuroscience, 1994.

That expert, Richard Ebstein: Natalie Angier, "Variant Gene Tied to a Love of New Thrills," *New York Times* Jan. 2, 1996, page 1.

98 *Several months later, however:* Natalie Angier, "Maybe It's Not a Gene Behind a Person's Thrill-seeking Ways," *New York Times*, Nov. 1, 1996, page 22.

101 *"[The] remodeling of the brain:* Linda Patia Spear, "Neurobehavioral Changes in Adolescence," *Current Directions in Psychological Science* 9, no. 4 (August 2000).

102 *In his lab, Lane and his colleagues*: Scott D. Lane and Don R. Cherek, "Risk taking by adolescents with maladaptive behavior histories," *Experimental and Clinical Psychopharmacology* 9, no. 1 (2001): 74–82.

Chapter 8 CORNY JOKES AND COGNITION

112 *As Fischer outlined this process:* Geraldine Dawson and Kurt W. Fischer, eds., *Human Behavior and the Developing Brain*, (The Guilford Press, 1994), chapter 1, pages 3–66.

114 *In March 2001, after Charles Andy Williams:* Daniel R. Weinberger, "A Brain Too Young for Good Judgment," *New York Times*, March 10, 2001.

Perhaps predictably: "Anatomy of a Teenage Shooting," *New York Times*, March 13, 2001.

116 *To define maturity, the researchers:* Laurence Steinberg and Elizabeth Cauffman, "Maturity of judgment in adolescence: Psychosocial factors in adolescent decision making," *Law and Human Behavior* 20. no. 3 (1996).

120 *While many have considered:* Natalie Angier, "Why We're So Nice: We're Wired to Cooperate," *New York Times*, July 23, 2002.

121 *One of the country's most respected neuroscientists:* "Impairment of social and moral behavior related to early damage in human prefrontal cortex," Steven W. Anderson, Antoino Bechara, Hanna Damasio, Daniel Tranel and Antonio R. Damasio, *Nature Neuroscience* 2, no. 11. (Nov. 1999): pages 1032–37.

Chapter 9 SWEPT AWAY

126 *Andrew Sullivan, a writer:* Andrew Sullivan, "The He Hormone," *New York Times Magazine*, April 2, 2000, pages 46–57.

127 *Not long ago, Liz Susman:* Jordan W. Finkelstein, Elizabeth J. Susman, Vernon M. Chinchilli, Susan J. Kunselman, M. Rose D'Arcangelo, Jacqueline Schwab, Laurence M. Demers, Lynn S. Liben, Georgia Lookingbill, Howard E. Kulin, "Estrogen or testosterone increases self-reported aggressive behaviors in hypogonadal adolescents," *Journal of Clinical Endocrinology and Metabolism* 82, no. 8 (1997).

129 *In their studies, Susman and her colleagues:* Jordon W. Finkelstein, Elisabeth J. Susman, Vernon M. Chinchilli, M. Rose D'Archangelo, Susan J. Kunselman, Jacqueline Schwab, Laurence M. Demers, Lynn S. Liben and Howard E. Kulin, "Effects of Estrogen or Testosterone on Self-Reported Sexual Responses and Behaviors in Hypogonadal Adolescents," *Journal of Clinical Endocrinology and Metabolism* 83, no. 7 (1998): 2281–85.

132 *Art Arnold, along with:* Fernando Nottebohm and A. Arnold, "Sexual dimorphism in vocal control areas of the songbird brain," *Science* 194 (1976): 211–13.

And if you give a shot of testosterone: Judy Cameron, "Effects of Sex Hormones on Brain Development," in *Handbook of Developmental Cognitive Neuroscience*, ed. C. A. Nelson and M. Luciana (MIT Press, 2001).

133 *In her experiments, Becker has shown:* Jill B. Becker, "Sex Differences in the Effects of Estrogen on Striatal Dopamine Activity and Sensorimotor Function," National Institutes of Health, Gender and Pain Abstracts, April 1998.

134 *As Deborah Blum lays out:* Deborah Blum, *Sex on the Brain, the Biological Differences Between Men and Women*, (Viking Penguin, 1997), pages 37–63.

136 *Giedd found that the amygdala:* Jay N. Giedd, A. Catherine Vaituzis, Susan D. Hamburger, Nicholas Lange, Jagath C. Rajapakse, Deborah Kaysen, Yolanda C. Vauss, and Judith L. Rapoport, "Quantitative MRI of the temporal lobe, amygdala and hippocampus in normal human development: Ages 4–18 years," *Journal of Comparative Neurology* 366 (1996): 223–30.

138 *Sheri Berenbaum of Penn State University:* Sheri A. Berenbaum, "Effects of early androgens on sex-typed activities and interests in adolescents with congenital adrenal hyperplasia," *Hormones and Behavior* 35 (1999): 102–10.

143 *Working with . . . Bradley Cooke, Breedlove:* Bradley M. Cooke, Winyoo Chowanadisai, and S. Marc Breedlove, "Post-weaning social isolation of male rats reduces the volume of the medial amygdala and leads to deficits in adults sexual behavior," *Behavioural Brain Research*, (2000).

Chapter 10 · THE NEURONS OF LOVE

149 *Over the past several years, Bruce Arnow:* B. A Arnow, J. E. Desmond, L. L. Banner, G. H. Glover, M. L. Polan, T. F. Lue, S. W Atlas, "Brain activation and sexual arousal in healthy, heterosexual males," *Brain* 125 (2002): 1014–23.

150 *There's been a rash of studies lately:* Winifred Gallagher, "Young Love: The Good, the Bad and the Educational," *New York Times*, Nov. 13, 2001, section F, page 6.

153 *Analyzing studies of older adolescents:* Martha K. McClintock and Gilbert Herdt, "Rethinking Puberty: The Development of Sexual Attraction," *Human Development* 5, no. 6 (December 1996).

155 *More recently, McClintock:* Nicholas Wade, "Scent of a Man is Linked to a Woman's Selection," *New York Times*, Jan 22, 2002, section F, page 2.

As Deborah Blum points out in her book: Deborah Blum, *Sex on the Brain, the Biological Differences Between Men and Women* (Viking Penguin, 1997), page 12.

Chapter 11 WAKE UP! IT'S NOON

159 *Over the past several years, Carskadon:* Mary A. Carskadon and William C. Dement, "Multiple sleep latency tests during constant routine," *Sleep* 15, no. 5 (1992): 396–99; Mary A. Carskadon, Cecilia Vieira and Christine Acebo, "Association between puberty and delayed phase preference," *Sleep* 16 no. 3 (1993): 258–62; Mary A. Carskadon, Amy R. Wolfson, Christine Acebo, Orna Tzischinsky, and Ronald Seifer, "Adolescent sleep patterns, circadian timing and sleepiness at a transition to early school days," *Sleep* 21, no. 8 (1998).

161 *Other studies by Wolfson and Carskadon:* Amy R. Wolfson and Mary A. Carskadon, "Sleep schedules and daytime functioning in adolescents" *Child Development*, 69, no. 4 (August 1998): pages 875–87.

164 *In his book Children's Dreaming and the Development of Consciousness*: David Foulkes (Harvard University Press, 1999), page 55.

166 *Several years ago, there was an experiment:* Erica Goode, "Rats May Dream, It Seems, of Their Days at the Maze," *New York Times*, Jan. 25, 2001, page 1.

Chapter 12 FALLING OFF THE TRACKS

175 *One study found that:* Sandra A. Brown, Susan F. Tapert, Eric Granholm, and Dean C. Delis, "Neurocognitive Functioning of Adolescents: Effects of Protracted Alcohol Use," *Alcoholism: Clinical and Experimental Research* 24, no. 2 (2000): page 164–71.

And additional studies by Brown: "fMRI Measurement of brain dysfunction in alcohol-dependent young women," *Alcoholism: Clinical and Experimental Research* 25, no. 2 (February 2001).

176 *According to federal surveys:* "Monitoring the Future," the University of Michigan's Institute for Social Research and the National Institute on Drug Abuse, 2000.

Other studies show: "How to Manage Teen Drinking," *Time*, June 18, 2001, page 42–44.

Most experts say solutions: Barry Stanton, "Harrison Players Still Just Don't Get It," *Journal News*, Sept. 25, 2002, page 1C; Jayne J. Feld, "Teen Use of Alcohol on the Rise, Experts Say," *Journal News*, Sept. 28, 2002, page 1A; David Novich, Karen Meaney and Meryl Harris, "200 Students Drunk at Dance," *Journal News*, Sept. 26, 2002, page 1A; Jane Gross, "Teenagers'

Binge Leads Scarsdale to Painful Self-Reflection," *New York Times*, Oct. 8, 2002, section B, page 1.

178 Studies from Swartzwelder's lab: H.S. Swartzwelder, W. A. Wilson, and M. I. Tayyeb, "Differential sensitivity of NMDA receptor-mediated synaptic potentials to ethanol in immature vs. mature hippocampus," *Alcoholism: Clinical Experimental Research* 19 (1995): 320–23; H. S. Swartzwelder, W. A. Wilson, and M. I. Tayyeb, "Age-dependent inhibition of long-term potentiation by ethanol in immature versus mature hippocampus," *Alcoholism: Clinical Experimental Research* 19 (1995): 1480–85.

179 And while a little calcium: M. A. Prendergast, B. R. Harris, S. Mayer, J. A. Blanchard, D. A. Gibson, J. M. Littleton "In vitro effects of ethanol withdrawal and spermidine on viability of hippocampus from the male and female rat," *Alcoholism: Clinical and Experimental Research* 24 (2000): 1855–61.

180 One brain-imaging study at the University of Pittsburgh: Bernice Wuethrich, "Getting Stupid," *Discover*, March 2001, pages 55–63.

184 But two years ago, a long-term study found: Jeffrey G. Johnson, Patricia Cohen, Daniel S. Pine, Donald Klein, Stephanie Kasen, and Judith S. Brook, "Association between cigarette smoking and anxiety disorders during adolescence and early adulthood," *Journal of the American Medical Association* 284 (Nov. 8, 2000): 2348–51.

186 In one study, Slotkin found: Theodore A. Slotkin, "Nicotine and the adolescent brain: Insights from an animal model," *Neurotoxicology and Teratology* 24 (2002): 369–84.

Chapter 13 INTO ANOTHER WORLD

189 But John Forbes Nash Jr., the Nobel Prize–winning mathematician: Sylvia Nasar, *A Beautiful Mind: A Biography of John Forbes Nash Jr.* (Simon & Schuster, 1998).

190 A human teenager's blood pressure: L. P. Spear, "The adolescent brain and age-related behavioral manifestations," *Neuroscience and Biobehaviorial Reviews* 24 (2000): 417–63.

192 But those who develop schizophrenia: Paul M. Thompson, Christine Vidal, Jay N. Giedd, Peter Gochman, Jonathan Blumenthal, Robert Nicolson, Arthur W. Toga, and Judith Rapoport, "Mapping adolescent brain change reveals dynamic wave of accelerated gray matter loss in very early-onset schizophrenia," *Proceedings of the National Academy of Science*, 98, no. 20 (Sept. 25, 2001).

192 *David Lewis, a neuroscientist at the University of Pittsburgh:* David A. Lewis, and Pat Levitt, "Schizophrenia as a Disorder of Neurodevelopment," *Annual Review of Neuroscience* 25 (2002): 409–32.

Chapter 14 COMING OF AGE

209 *Even the U.S. Navy:* Denise Grady, "Sleep Is One Thing Missing in Busy Teenage Lives," *New York Times*, Nov. 5, 2002, section F, page 5.

Index